WHERE SCIENCE WENT WRONG

TRACKING FOUR CENTURIES OF MISCONCEPTIONS

Peter Bros

1997

INFORMATION
PHONE 1-703-971-9141
FAX 1-703-971-1628
info@copernican-series.com
http://www.copernican-series.com

PUBLISHED BY FINANCIAL BOOK PARTNERS

PRINTED IN THE UNITED STATES OF AMERICA

ISBN 0-9627769-7-1

Designed by Lemieux Creative Enterprises

WHERE SCIENCE WENT WRONG
TRACKING FIVE CENTURIES OF MISCONCEPTIONS

Peter Bros

i FORWARD

1 1. A PROPERTY THAT PRODUCES MOTION: THE NONEXPLANATION FOR FALLING OBJECTS

21 2. THE OCCULT NOTION OF MOTION

41 3. MASS/GRAVITY IS BASED ON THE OCCULT NOTION OF STRAIGHT- LINE MOTION

59 4. THE SOLAR SYSTEM DOES NOT OPERATE ON THE BASIS OF HISTORICAL FORCES

77 5. TURNING BELIEF INTO FACT

97 6. WE CAN SEE LIGHT FROM STARS AT THE END OF THE UNIVERSE AND THE BEGINNING OF TIME AND FIBBER MCGEE AND MOLLY ARE OUT THERE SOMEWHERE

117 7. FOUR HUNDRED YEARS OF PROGRESS: FROM BEING AT THE CENTER OF THE UNIVERSE TO BEING ISOLATED AND ALONE IN THE UNIVERSE

151 8. THE EARTH IS BILLIONS OF YEARS OLD

179 9. THE NONEXISTENT NEUTRON

The Copernican Series

Producing a consistent picture of physical reality

How we analyze physical reality is the subject of
Volume 1, ***At the Gates of the Citadel: The Subjugation of Modern Science***

What life is and how it evolves is covered in
Volume 2, ***The Cooling Continuum: The Rise and Fall of Species on Earth***

How the universe operates is covered in
Volume 3, ***Atoms, Stars and Minds: Synthesizing an Elementary Particle That Comprehends Itself***

What the mind is and how it operates is covered in
Volume 4, ***The Model Mind: What the Mind Is and How It Works***

How the atmosphere transports heat is covered in
Volume 5, ***How the Weather Really Works!***

Light, electricity and magnetism are unified in
Volume 6, ***Light: Replacing Three Centuries of Misconceptions***

Where it all went south is set forth in
Volume 7, ***Where Science Went Wrong: Tracking Five Centuries of Misconceptions***

How the mind drives behavior is covered in
Volume 8, ***Human Nature: How the Mind Generates Behavior***

The mind body connection is described in
Volume 9, ***How the Body Really Works!***

The Ideography, where it all came from is found in
Volume 10, ***Transparent Realities***

Companion Titles

***Let's Talk Flying Saucer*: How Crackpot Ideas Are Blinding Us to Reality and Leading Us to Extinction**

***Production Based Prosperity*: A Challenge for the Millennial Generation**

***The Credit Union*: Its Position in the Consumer Financial Marketplace**

FORWARD

The only conclusion concerning measurable physical phenomena that Empirical Science and its precursor philosophy have gotten right since records of knowledge have been kept is that the Earth moves around the sun, and it is only in the last five hundred years that it has willingly accepted this conclusion.

Since that observation, delivered to the world on the deathbed of its proponent Copernicus who feared that its earlier delivery would be contemporary with his untimely death, Empirical Science has promulgated a series of misconceptions concerning measurable and nonmeasurable physical phenomena that, compounded one on top the other, has produced a picture of physical reality so absurd that it would be laughable were it not for the ignorance the picture casts over the human progress for which Empirical Science claims responsibility.

The number of misconceptions is legion, filling every field of knowledge. *Where Science Went Wrong* includes the major misconceptions surrounding concepts of force and motion. The misperceptions that led to the misconceptions, the pathways to delusion, are set out in detail.

Except for the final chapter, The Nonexistent Neutron, *Where Science Went Wrong* does not provide answers to the questions about physical reality that underlie each misconception. For empirical praetorians charged with wielding the ultimate shield against scientific apostasy, the plea that what can't be replaced can't be reproached, the answers appear in the first six volumes of The Copernican Series.

A NOTE ON SCIENTIFIC COMITY

One of the basic tenets of Empirical Science is said to be the need for intelligent people to sit down calmly and discuss on a reasonable basis and in a respectful manner the issues underlying the vast human enterprise that is the search for knowledge in the universe. Because science claims to underlie all human progress, we are instructed that invective, sarcasm and above all name-calling are to be left at the door of the meeting hall so that all present may participate in the improvement of science and the human condition in a quiet and orderly fashion.

This crock of crap, meant to silence those who don't agree with the consensus, is similar to the incantations mumbled by ancient priests, hands held high, forefingers crossed over tortured sacrifices to ward off evil spirits, evil spirits in science being original thought, the tortured sacrifices the remains of all who disagree with the consensus.

Science regularly refuses space in its journals to ideas that oppose the consensus, editors, in fact, having been known to publish lists of subjects which are off-limits for scientific discussion. Observatory facilities are regularly denied for queries into areas accepted as fact, settled interpretations for observations such as the red shift in light when there is no coherent explanations for what light is to start with.

In the era of big machine science, the big machines are off-limits unless they can be used to prove hypothetical facts that are predicted from scientific facts, hypotheses that have been mistaken for reality.

Worse, students can't get doctorates expounding unorthodox ideas, teachers can't get tenure for expressing nonconsensual hypotheses, and professors, should they oppose the established canon either in class or in print, will be hounded from their posts, lose grants, be kicked out of professional societies, their name blackened, cast from the lists, even banned from open-minded establishments of higher learning.

When practical scientists working in subsidized laboratories uncover information that opposes the consensus, they are smeared in the name of the consensus as being bought, tainted by the commercial interests that underwrite their efforts although all advancements in chemistry and physics from Edison through Rutherford to the present, unless the product of pure happenstance, have been subsidized by commercial (or military) interests. Consensus controlled testing laboratories refuse to certify measurable phenomena that oppose the consensus conclusions about what concepts control our picture of reality in fear of losing accreditation.

Work in areas considered to be unproductive, or settled, that produces results that oppose scientific convention, even when adhering to strict empirical standards as promulgated by the scientific establishment, is made the butt of jokes and derision. Scientific meetings, where results are supposed to be objectively evaluated, become the forum for such infantile invective that the fear of ridicule prevails, forcing some to hold sessions only for the safe few, others to forego communicating results altogether, or even from attempting to obtain results to start with.

The use of scientific comity for protection against criticism of process is equally ludicrous. The two hallmarks of the so-called objective scientific method are jokes. Peer review in an atmosphere where consensus rules is used simply to exclude unorthodox points of few, while specialization turns every group of peers into a miniature in-group, speaking a language only group members understand. As a result, any agreed upon absurdity is acceptable so long as it's presented empirically and its underlying hypotheses don't conflict with basic scientific conclusions about reality, interpretations of fact encoded as law for easy enforcement.

Even more ludicrous is the claim of duplication of experimental results. Not mentioning for a moment that interpretations of fact have nothing to do with the facts they interpret so that how experimental results are viewed rests solely on the consensus opinion as to what those results mean, the only reason to duplicate an experiment is to disprove those that conflict with the consensus. In the era of big machine science, and costly experiments, verification occurs in the experimental design itself, and if the verification doesn't occur in the execution of that design, the experiment is redesigned until verification does occur. There is no question of repeating experiments that reinforce the consensus by appearing to find the results its hypotheses predict. The only experiments that are conducted on a repeated basis are easily performed curiosities whose interpretations can be used to promulgate the scientific consensus and those rigged to debunk experiments that have produced results that oppose the consensus.

The reason that Empirical Science needs to defend to the death the delusional concepts it creates is that, like religion, it is based on a set of revealed truths, laws, which have to be believed before one can become a participant in, and be protected by, the resulting belief system and benefit from the structure that has been crafted around it.

The core delusional concept of Empirical Science is the belief that concepts can be proven to be fact by hoking up some sort of methodology, basing the belief, for heaven's sake, on the ancient practice of casting a star speckled, robed arm to the sky, mumbling some incoherent incantations, and claiming to be able to predict the future by reference to the lights in the firmament.

The delusion underlying a process that claims as fact beliefs that predict other facts and then calls the method scientific is self-evident.

Because facts are facts and beliefs are beliefs, producing a procedure that turns beliefs into facts involves all sorts of mumbo jumbo created to confuse nonbelievers, impress them as to the superior knowledge status of the believers, and as a result, gull the nonbelievers into forking over additional funding to support the believers' lifestyles while the believers sit around basking in the glory, the elegance, *The Truth* of their beliefs. Cries for funding basic research serve only to perpetuate the scam, implying a vast body of omnipotent scientific knowledge out of the public eye that is responsible for technological advancement, the claim that the pile of delusional concepts, occult beliefs such as gravity as a property of matter for why released objects move, angular momentum as conserved in closed systems for why planets rotate or swirling masses of condensing gas for why planets orbit, are responsible for microwave ovens and television sets.

When the obvious misconceptions are questioned by the public, the things of science fiction like black holes, antimatter, space warps, wormholes or time travel, or the things beyond science fiction like strings, singularities, supersymmetry, supergravity, electroweak forces and quantum chromodynamics with its top and bottom quarks, Empirical Science falls back on its methodology, claiming "We didn't think these profound observations up, our objective methodology led us to them inexorably!" -- the claim, espoused by Empirical Science's religious twin that "the Lord works in mysterious ways!"

As noted in the first chapter of the first volume of The Copernican Series, not even the ultimate authority can turn a belief into a fact, and a scientific community that believes its beliefs to be facts will be living a delusion, an accurate statement of current scientific thought.

As the volumes of The Copernican Series illustrate, the only authority is the authority of ideas to actually explain on a consistent basis how facts in reality work together to produce the reality in which we find ourselves.

Currently, science has no ideas that explain facts coherently, and none of the ideas that explain some of the facts reasonably well are consistent with one another.

As a result we have a blustering old science with its silly beliefs and gross misconceptions protecting itself with the shield of comity.

COMMENTS ON THE MATERIAL

Especially in the first four chapters dealing with force and motion, and the fifth chapter, dealing with the scientific procedure derived from the erroneous belief that questions of force and motion had been resolved, but generally throughout *Where Science Went Wrong*, there are various depictions of the facts surrounding Galileo's claim that he proved that mathematics could turn beliefs into facts and Newton's claim that he had demonstrated the source of the force that make objects fall.

That's because these two pivotal points are where science went wrong!

What appears, at first glance, to be duplications of explanation are not when the context of the explanations is taken into consideration. The mass/gravity concept is an interrelated group of misconceptions that has to be peeled like an onion, one layer at a time, until it can be laid bare as a mistake of process.

And that is how this, the seventh volume of The Copernican Series differs from the first six volumes. The dominate theories about reality that are, today, accepted as fact, are not worth much explication because, priding themselves on being counterintuitive and basing their validity on the nonexistence of a mind capable of both reason and rational analysis, they make little attempt to address the immediate questions about how physical reality operates, the answers to which are of utmost importance if we are to extend our range of survivability in that physical reality. The real questions that concern us, or at least should concern us, what is energy, what causes motion, have long since been abandoned in favor of nonanswers that use words as a substitute for comprehension.

The task of explaining not only how physical reality operates, but how we can know that we are actually finding out how physical reality operates, how the mind forms pictures of that reality and how the pictures the mind forms relate to reality, required a lot of ink without wasting more ink on subject matter that can be readily found in comic books, easy readers and on television.

However, the limitation of this approach resulted in a lot of explanations left standing that are taken blindly as proof for things that have no basis in reality. Even though providing a mechanical explanation for Young's two-slit experiment is alone sufficient to knock the underpinnings out of quantum mechanics, and the failure of interference patterns to prove that light is a wave explains the nonresults of the nineteenth century Michelson and Morley experiment, thereby negating ruminations based on the so-called Fitzgerald contraction, the predecessor, whether acknowledged or not, of relativity, these two dominant twentieth century belief systems have converged to produce a preposterous picture of the nature of matter as a result of the discovery of nuclear energy.

Nuclear energy is such an anomaly in the orderly development of any technology that the window it produces on the nature of matter should not be shaded by unfounded theories that do not account for the nature of energy and use words as substitutes for comprehending force and motion.

The experiments that led to the production of nuclear energy provide vast amounts of information and that information needs to be put in context.

However, it's a chapter's worth, not a book's worth.

The failure to understand the nature of energy, though, and the resulting failure to explain force and motion are precisely the reason we live in a bubble of ignorance dealing with physical reality.

As a result, the subject of the nonexistent neutron, the neutron being the word used as a substitute for understanding nuclear energy, makes an ideal conclusion to a volume on how that bubble came to exist, how we came to live with the delusion that we know everything when we know absolutely nothing.

Trying to burst a bubble that no one recognizes exists presented an interesting challenge. Like a tangled ball of string, it's hard to unravel a bubble of unfounded beliefs when its core, the theory around which the ad hoc threads of thought are tangled, is uncritically accepted as a statement of factual reality. When discrepancies in the core belief are uncovered, interpretations are woven out of thin air and another layer added to the ball making it increasingly difficult to get at what the actual nature of reality is.

For instance, relating the red shift of stellar spectra to distance might well be a fun notion so long as one keeps in mind that the physical basis of the red shift is unknown, that it is impossible to visualize how multiple waves of light can travel together, that we have no way of knowing whether the red or blue wavelengths are the shorter wavelengths, that there is no physical evidence for a medium through which light can travel and therefore no basis to analogize light to a train whistle blowing in the wind that would therefore shift, that with no medium there is no physical basis for claiming light to be a wave and thus capable of shifting, and that, with light being an imponderable, there's nothing to shift!

It might be a fun notion so long as it remains a notion.

However, when the abysmal ignorance about the light that gives rise to the red shift is further obscured by the conclusion that the universe is expanding because the magnitude of the red shift is proportional to distance, the fun notion becomes a search for proof that the universe is expanding.

In search of the proof that the universe, space, nothing, is actually expanding, an additional layer of string is added to the ball by dreaming up the fantasy that if the universe is expanding, and the universe is made up of all those fiery stars, then the universe must be the result of an explosion, fire being the result of explosions on Earth.

Moving rapidly from fantasy to fact, we have replaced an ignorant interpretation of light with a big explosion, something we are all aware of in physical reality. We mistake the reality we know for the fantasy we have made up.

What do explosions produce but light, and while light is an imponderable, explosions also produce energy, radiation, the empirical definition of which is rays of energy, rays being what come out of the center of something, with all explosions having centers.

And because the universe is the only thing around, the radiation produced by the explosion has nowhere to go so it must still be around, existing as a remnant, the background radiation left over from the explosion.

Not knowing anything about the nature of energy, we don't have to concern ourselves with how this background radiation can be left over as a remnant from some sort of explosion that now creates matter rather than, as an explosion does on Earth, destroys matter. As a reult, the validity of the ball of string, now no longer a bubble, but a reality, can actually be verified by finding the background radiation that is left over from the universe exploding into existence.

And, with only one universe and one explosion, the background radiation would be left over uniformly everywhere.

When the radio telescope was invented, it was found that it picked up a background static of about three percent no matter in which direction the dish antennae of the radio telescope was pointed. A little bit of mathematical prestidigitation quickly demonstrated that three percent would be the exact amount of background radiation that would have been left over from the explosion, now a Big Bang, that brought the universe into existence.

One completed ball of string!

But science proceeds as a result of brilliant observations and someone observed that while the remnant of the Big Bang, the background radiation, was thought to be uniformly distributed throughout the universe, the visible remnants of the Big Bang, the galaxies, were not.

Thus, there was something wrong with the ball of string. It was in danger of becoming a big ball of gas!

Unless the background radiation isn't uniformly distributed after all.

The only way to find out whether the background radiation was uniformly distributed or not was to get closer to it by sending a rocket to measure it above the atmosphere, the assumption being that uneven radiation in space was being uniformly distributed by passing through the atmosphere (unlike starlight, which is uniform until it passes through the atmosphere after which it becomes uneven, twinkling starlight!)

The COBE satellite was launched to determine if the background radiation was indeed uniformly distributed throughout the universe. Finding that it wasn't, that there were actually ripples in the cosmos, whatever the cosmos is, the interpretation of the red shift as related to distance was considered proven beyond a shadow of doubt, a perfect ball of string.

Of course, there are current nuclear processes going on inside the Earth which presumably produce the same radiation that would be left over from the Big Bang. No matter which way the dish of the radio telescope was pointed to pick up a uniform three percent radiation, the dish's backside would be receiving the steady flow of radiation from the surface of the Earth.

And when the COBE satellite is sent on the other side of the atmosphere, the atmosphere would serve to disrupt the regular measurement of this radiation turning its uniform measurement irregular!

But then if reality were considered, the measurement of the radiation couldn't be used to justify the belief in the Big Bang, Empirical Science's creation myth!

The failure to understand energy, the source of force and motion, nullifies all conclusions related to force and motion, and specifically the imaginary movement the Big Bang was manufactured to explain.

When we wrap our ignorance up into balls of string which grow larger and larger as ad hoc explanations are strung together, we lose the ability to question our basic ignorance, accepting ignorance, nonknowledge, the lack of any coherent explanation as a shining ball of gas, the most perfect and precise of explanations.

The use of atomic clocks to prove relativity's contraction of time is the pre-eminent example of a physical effect based totally on the nature of energy being interpreted with a total ignorance of energy, here that the energy even exists. If we don't know that the Earth is an emitter, that it is producing a constant outflow of energy, the same outflow that is mistaken as the background radiation proving the Big Bang, and that that outflow produces a field in which matter exists, we have no knowledge that the field in which the atomic clock exists controls the rate at which the atomic clock itself emits, the rate of emission being the clock's measurement of time.

When the atomic clock is on the ground, it is in the field of temperature that exists at the surface of the Earth. Just like a piece of tin glows hot in accordance with how close it is to the fire, so to do atomic emitters emit in accordance with the field that they occupy.

As the Earth's emission field expands out and away from its surface, it diminishes, just as gravity is measured to diminish, inversely with the square of the distance from the surface of the Earth.

Thus, the field several hundred miles above the Earth is substantially less than the field at the surface of the Earth, and the rate of emission of the atomic matter that makes up the clock is substantially less. With the rate of emission reduced, the time the emissions measure appears to slow down.

But, because we believe that time slows down as speed increases, we interpret the reduction in the rate of emissions of the atomic clock as a result of the increased speed of the vehicle carrying the clock rather than the reduced field the clock occupies, because not knowing anything about energy, we aren't even aware of the field, let alone its reduction, or how matter acts when a field is reduced.

The mistake involving the creation of the nonexistent neutron outlined in the last chapter of *Where Science Went Wrong* highlights the extent to which the lack of awareness of the nature of energy can sink us further and further into an ignorance that can only lead to our demise.

When we don't ask the correct questions, we cannot expect

to get correct answers. The failure to phrase questions about physical reality correctly led to the Galilean confusion, it led to the Newtonian confusion and it has led to our own confusion.

But with respect to energy, and its effects, no one has even asked the question!

The question has always been, what makes things move?

No one has asked the question because the cause of motion, and thus momentum, and thus force has been assumed away, whether it is the motion of the nonexistent neutron or the motion of the planets.

By assuming away motion, we have reversed reality, looking for energy where it does not exist and ignoring energy where it does exist.

This reversal of reality is at the basis of The Copernican Series.

Copernicus, who Empirical Science claims uncovered no scientific fact, reversed reality, providing a clear picture of the solar system by placing the Earth orbiting the sun, showing in the process how the planets move with respect to the sun.

The Copernican Series uncovers no scientific fact. It merely requires that current motion require current force and explains what that current force is and how it produces current motion, showing in the process how the solar system moves with respect to the galaxy.

The error Copernicus corrected dealt with the vision of our position in the universe.

The error with respect to motion deals with our continued presence in the universe, whether we continue to believe that our names, our strings of words, our mindless laws define the universe, or whether we can ask the simple questions that will allow us to understand how the universe operates.

Thinking we know everything when we know nothing may elevate our status from that of savages to that of noble savages, but being totally ignorant of the nature of reality, we remain savages none-the-less.

When we look back on the simple shift in the point of view brought on by Copernicus, we can, with printing still in its infancy, identify few protagonists whose steadfast belief in unreality turned them into laughing stocks. Today, with preposterous absurdities being the substance of tabloids, science columns in the daily papers, trade and association journals and most of all, of the esteemed annals of the various learned societies, the subtle change in point of view, from saying nothing moves something to finding something moves something, will provide identifiable sources of amusement for years to come.

1. A PROPERTY THAT PRODUCES MOTION: THE NONEXPLANATION FOR FALLING OBJECTS

NEVER HAVING TO ASK THE QUESTION

When we see an object fall, we don't have to ask the question why it falls because we know the answer instinctively. It falls because of gravity. We probably learn what gravity is before we learn about God because we become aware of falling objects before we become aware of our own existence. We have to keep away from falling objects, and keep from becoming a falling object ourselves, if we want to continue to exist.

We know from our first thoughts why objects fall. They fall because they have weight.

We also know where objects fall. They fall down, toward the surface of the Earth.

When we become somewhat sophisticated, we can know something else. By measuring the rate at which objects fall, we can determine that objects accelerate when they fall, and they do so with the square of the distance over which they fall.

This tells us that whatever is making objects fall is diminishing inversely with the square of the distance from its source. Because objects fall toward the Earth, we can pretty much conclude that the source of the force has something to do with the Earth.

We know something else, too. We know that whatever the force is that is making objects fall, it is a current force. It is present in the here and now.

And, of course, we should be able to realize that because something is occurring in real time, that an object is accelerating while it is falling, that whatever is making it fall exists between the source and the object.

If it is doing something, then it must be something.

With all of these clues we might conclude that someone, somewhere would have long ago sat themselves down and figured out what was happening when a dropped object began to move.

Instead, we said nothing, that objects fall because of a force called gravity and gravity is a property of matter.

DEFINING THE FORCE AS A PROPERTY ASSUMES AWAY A RATIONAL EXPLANATION FOR IT

A property is a quality that defines or describes an object or substance. Gold is the color that identifies the matter we call gold. The color is a property that defines the substance. Another of its properties is softness or malleability. Steel has a property of hardness, ice a property of coldness.

Properties are physical characteristics we use to define classes of matter.

What does that have to do with falling objects? We don't pick up a piece of dirt or a rock, or point to a car or a house and say, look, see its property of gravity.

Causing motion is not an appropriate property of mater because it cannot be used to identify matter. Motion is something that is happening in reality. When something happens, we need a description that will tell us what is happening.

Gravity is not a property of matter, it is something that happens to matter. It can't be used to classify matter.

However, because things were moving in reality, we knew there was some force, invisible, that was causing the motion.

We were at a loss what that force was.

We therefore labeled the nonapparent, but necessary force that makes objects fall gravity so that we would have a word to use when we were talking about why objects fall.

That word contains no explanation about what the force is, how it is generated, how it moves from one place to another, how it acts on matter to make it move, or what it is doing when it is not acting.

Saying that gravity is a property of matter is saying absolutely nothing about the force that makes objects fall. It is a senseless statement, a meaningless phrase, it has no content, it is empty, the sugar cube at the core of modern empirical science.

How is it that we can be blind enough to create an explanation for the most basic force in our existence that makes no sense in reality, does not provide a physical explanation for what is causing the force or explain how that force operates, that is basically the assignment of a name, and place that explanation at the center of our understanding of the universe?

GALILEO'S FLIGHT OF FANTASY, BRAHE'S MATHEMATICAL MISPERCEPTION

The empirical fanatasy that is the mass/gravity concept is the result of a convergence of a flight of fancy on the part of Galileo, the man history credits as being most forceful in bringing down the old belief that the sun traveled around the Earth and of a misperception on the part of Brahe, the last man to defend the old belief to the death.

GALILEO'S FLIGHT OF FANCY -- PERPETUAL MOTION

In attempting to prove that the sun didn't move around the Earth, Galileo sought to prove that it was the rotation of the Earth on its axis that gave the sun the appearance of movement. However, if he were to explain the sun's appearance of movement on the basis of the Earth's rotation, then he had to explain the Earth's rotation.

Because the force that causes planetary rotation is, like gravity, a nonapparent force, Galileo decided to assume it away rather than explain it. Unlike gravity, which made objects fall on you and thus couldn't be readily assumed away, nobody had ever been killed by the force that caused the Earth to spin. Galileo therefore took the tack that if something had set the Earth spinning, it would continue to spin so long as there wasn't anything to stop it from spinning. As space was devoid of anything, there wasn't anything in space to keep the Earth from spinning.

This sort of mental delusion is beneath contempt.

Objects don't move as a result of a force from the past.

They move with respect to a current force.

We see something moving in space, we are too lazy to sit down and figure out what is causing it to move, so we reverse reality and say, well, duh, it's moving, isn't it, so it'll keep right on moving unless something stops it from moving. The only time we'll have to trouble the old noodle is if we measure something slowing down. Then we can look for what's slowing it down, even though we still don't know what's making it move!

As proof for his claim of the Earth's perpetual motion, Galileo rolled balls down an inclined plane and measured how they moved when they rolled off the end of the plane. A half a century earlier, a man named Tartaglia had gone out onto the proving grounds and, by trial and error, determined that for distance the optimal angle for shooting off a cannon was forty-five degrees. Galileo was able to predict the same measurement in the confines of his study by reference to the rolling balls.

Galileo then promulgated one of the greatest misconceptions in history, stating that, because by mathematics alone he was able to predict the optimum angle of Tartaglia's cannonballs, that mathematics could be used to confirm as fact that which could not be directly measured.

Carrying this misconception to its next logical level, Galileo claimed that by mathematically extrapolating from his rolling balls he could state with absolute certainty that, if he were able to construct a frictionless path around the Earth, then his balls, once imparted with motion from the inclined plane, would roll around that path forever!

Galileo called the quality that allowed matter to move perpetually in frictionless space and thus the Earth to perpetually rotate on its axis, inertia.

Qualities, of course, are nothing but properties. There is no question that the Earth rotates on its axis. Galileo was simply attempting to come up with an explanation why it was doing so. But inertia, like gravity, is just a word to describe the physical result of an unknown force acting on matter, and has no meaning in and of itself, provides no explanation for what is causing movement, and in fact, with the Earth's rotation, assumes that cause occurred some time in the past, from Galileo's perspective, when God created the universe.

The error Galileo made was to believe a concept, inertia, could, by a system of mathematical extrapolation, be proven to be fact. Concepts are not facts, and can never be proven to be fact. Concepts are only mistaken to be facts by the gullible and the gulled.

Mathematical extrapolation works fine in producing technology because failures, not working in reality, allow trial and error to pinpoint success. In science, mathematical extrapolation just produces nondisprovable hypotheses which, because they withstand the test of time, which means they can't be disproven, become facts. Instead of the measurable fact that objects come to rest with respect to the forces acting on them, we create a nondisprovable hypothesis, objects move unless a force acts on them, and because it can never be disproven, label it a fact, or better, a law, reversing reality in the process.

Mathematical extrapolation doesn't work when it's used to make up stuff like perpetual motion.

Inertia went out the window when Galileo was called upon for demonstrable proof of the Earth's rotation in reality. He turned to the tides for the proof of the Earth's rotation because the tides act precisely as if the Earth's rotation were causing them. However, realizing that the sloshing of the tides would have constituted friction that would have long since overcome his hypothetical inertia, Galileo abandoned this line of thought, leaving his fanciful inertia hanging in nondisprovable space, an occult piece of fruit ripe for plucking by the gullible and gulled.

BRAHE'S MISPERCEPTION THAT THE MOON SPEEDS UP AND SLOWS DOWN

Brahe, the last defender of the preCopernican system, perpetuated the misperception that planets can speed up and slow down based on the appearance of a change in the movement of the moon much as the preCopernicans based their belief that the Earth was the center of the universe based on the appearance of the sun moving around the Earth.

Assume for a moment that we are all on a ship sailing slowly across a peaceful ocean. The pace, in fact, is so peaceful that we think we are standing still, just as the preCopernicans thought that the Earth was standing still in an ocean of space.

Being somewhat bored, we decide to start lifeboat races back and forth beside the ship. During trial heats we find that the teams always take less time to go from stem to stern than they take to go from stern to stem. Thinking this might be a product of the rowers becoming tired, we reverse the race, starting them at the back of the boat rowing to the front, and then again to the back.

We still find that the teams always take less time going from stem to stern, from front to back, than from stern to stem.

The mathematical formula that expresses movement is time times distance equals rate. If the time varies over the same distance, then the rate has to be changing. However, if the rate doesn't change, as would be indicated by reversing the starting point for the rowers, then the change in time has to be the result of a change in distance.

When we saw the teams taking less time one way than the other, and concluded they were slowing down because they were tired, we were assuming that the distance didn't change, that the distance between the stem and stern was as it appeared, the same as the distance between the stern and the stem.

The only clue we had that the distance might be changing, that the ship might in fact be moving, was the fact that changing the starting point didn't affect the outcome. The fast leg was always the leg from stem to stern, the leg from the direction in which the ship might be moving.

Knowing the ship was moving would have allowed us to know that the trip from stem to stern covered less distance because the stern was moving toward the lifeboat while the lifeboat was moving toward it and that when the lifeboat was moving from stern to stem, the stem was moving in the same direction the lifeboat was moving and the lifeboat had to go the added distance to catch up with it.

Even if we couldn't know, from appearance, that the ship was moving, faced with the fact that different times were being produced by the same apparent rate would have forced us to conclude that the ship was actually moving.

This is the type of mathematical extrapolation that works!

Tycho Brahe didn't have the clue provided by something taking different times to move the same apparent distance depending on the direction of travel when he set about measuring the speed at which the moon was traveling. Because the moon was traveling in a circle, he reasonably assumed that the distance the moon was traveling in its orbit was always the same distance.

Thus, when he found it speeding up in the summer and slowing down in the winter, it never dawned on him that the change in speed wasn't a change in speed at all, but rather a change in distance resulting from the movement of the Earth in its orbit around the sun, and the movement of the sun in space.

He thought the Earth was sitting stationary in space!

How could Brahe believe anything other than what he measured, that the moon was speeding up and slowing down. The facts were right in front of his eyes, as they are in front of ours. All anyone has to do is go out and measure the time it takes for the moon to complete an orbit after the winter solstice, and compare it to the time it takes for the moon to complete an orbit after the summer solstice to find out it slows down in the winter as the planet moves toward the winter solstice and speeds up in the summer as the planet moves toward the summer solstice.

The distance the moon is traveling in its orbit can't vary simply because its orbit is a circle. A circle is defined by the distance to its center and thus the distance something travels when it is going in a circle can't vary

This is the opposite of mathematical extrapolation. Measure something and draw the first conclusion that comes to mind. For thousands of years we went out and measured the sun moving across the sky and concluded it moved around the Earth.

Why shouldn't we conclude the moon is speeding up and slowing down in its orbit?

Certainly not because it is absurd to believe that something we think is moving as a result of a historical force could regularly change that rate of motion.

Certainly not because there is no source for the energy required to speed the moon back up once it has slowed down.

Rational thought is eclipsed by the measurable fact that the moon speeds up and slows down from season to season.

However, if the sun were in fact moving through space, then the distance the moon traveled in its orbit would be shortened by the distance the sun traveled while the Earth was moving away from the direction of the sun's motion, and the distance the moon traveled in its orbit would be increased by the distance the sun traveled while the Earth was moving toward the direction of the sun's motion.

This is in fact what happens as the Earth moves around the sun. Like the lifeboat example, as the Earth moves toward the summer solstice, it is moving away from the direction of the sun's motion, and thus the distance it travels, and the distance the moon also travels, is shorter than the distance the moon travels when the Earth has to catch up with the sun's motion as the Earth travels toward the winter solstice.

In fact, the differential in times allows us to calculate the speed of the sun!

So Brahe, instead of using mathematics as a tool of reason, used it to reverse reality, leading him to conclude that the Moon sped up and slowed down in its orbit.

So what?

KEPLER COMPOUNDS BRAHE'S MISPERCEPTION

Kepler carried Brahe's misperception into his measurements of the orbit of Mars.

Mars, which has the most irregular appearing orbit of any of the planets because the Earth's motion makes it appear to travel backwards at times is also subject to the same misperception of measurement as the moon. While Mars appears to speed up and slow down in its orbit, it in fact is actually moving at a uniform rate with the time difference being the result of it traveling different distances due to the sun's motion.

Kepler had computed the speeds of the known planets and found that the further they were placed from the sun, the slower they moved. In addition, he had discovered that the orbits of the planets were not circular, something that even Galileo hadn't suspected. Kepler found that the orbits of the planets were ellipses. We can visualize these orbits if we think of the orbits being circular but for the fact that the sun is moving slowly through space. As the sun moves through space, it is dragging the planets slightly behind it so that their orbits are slightly stretched out, with one portion of the orbit being closer to the sun than the opposite portion.

Because of his measurement of the planets' speeds becoming slower with their distance from the sun, Kepler took the misperception that Mars sped up and slowed down in its orbit just like the moon did and created a mental image of the planets speeding up and slowing down in their orbits in concert with their distance from the sun.

If anyone wanted proof for the statement, all he had to do was look at the different speeds of the planets in their orbits. The further they were from the sun, the slower they went and the closer they were to the sun, the faster they went.

It is one thing for different planets to go different speeds at different distances from the sun and quite another for one planet to speed up and slow down in the same orbit, but in his desire to produce a universal mathematical law, Kepler only had to point to the variable speed of the moon as proof that planets could speed up and slow down.

The mathematical law that Kepler created was totally without basis in reality. He said that a planet's speed was proportional to its distance from the sun. The closer it came to the sun, the faster it moved. The further it went from the sun, the slower it moved. This resulted in the fact, easily demonstrated mathematically on paper because the orbits were ellipses, that any two equal times in a planet's orbit would produce areas swept out by the planet's movement that were also equal.

This law, that a planet sweeps out equal areas in equal times is disprovable simply by sticking a stake in the ground and timing the movement of the sun. The sun reaches its northern most point during the summer solstice and its southern most point at the winter solstice. These two points are on opposite sides of the Earth's orbit so that, if Kepler's laws were correct, if the Earth were sweeping out equal areas in equal times, the times it takes for the Earth to move between these two points in its orbit, each describing exactly one half of the area of its orbit, should be equal. However, it takes about forty more hours for the Earth to move between the summer and winter solstices than it does for the Earth to move between the winter and summer solstices because the sun is moving in the direction of the winter solstice.

If Kepler's law is so easily disproven, how come it is still a law? Laws, by definition are correct, and being the consensus of the scientific community, reality must fall if it fails to agree. Empirical science, claiming to base itself in reality by measuring facts, actually just creates nondisprovable hypotheses which then become incontrovertible facts, facts which, because they are true regardless of reality, dictate reality itself.

Conforming planetary motion to Kepler's laws actually increases the time differential between opposite points in the Earth orbit to seventy two hours, disproving itself by its own operation!

One mathematical relationship inherent in Kepler's laws is the fact that an ellipse is a conic section which, because it is circular, has an area that is described by the square of its variable radius. The perceptive can quickly see that, by equating the areas swept out by a planet's passage with the time of the planet's passage, Kepler had inadvertently dictated that the change in speed that occurs as a planet speeds up and slows down had to be related to the square of its radius from the focus of the ellipse, the sun, because the area of a circle is determined by a constant, pi, and the square of its radius.

By making up a law that tied the speed of a planet and its distance from the sun to the area swept out by its passage, Kepler had forced the planet to not only speed up and slow down in its orbit, something it wasn't doing in the first place, but to speed up and slow down in a specific manner, at rates that were proportional to the square of its distance from the sun.

Thus, Kepler inadvertently created the same relationship for a planet supposedly accelerating toward the sun as existed for a rock falling to the surface of the Earth. The reason for the relationship wasn't the same, with a rock falling in an expanding sphere of force which diminished inversely with the square of its distance from the direction toward which the rock fell, and the planets doing so by dint of complying with a law equating the areas they swept out, but by the time Newton finished with the relationship, reason wasn't an issue.

NEWTON'S ALREADY DONE IT!

Robert Hooke, Christopher Wren and Edmund Halley, the early movers of the British Royal Society, sat around one day speculating that it was the matter in the sun that made the planets move in their orbits. Kepler had posited that it was something about the sun's rays that caused the planets to move, but this was easily discarded as a hypothesis because the moon was orbiting the Earth and the Earth didn't look anything like the sun.

As the Earth wasn't a ball of fire, and the sun was, there must be something else the two had in common that made objects fall toward them. Well, they must both be made up of matter. If it could be mathematically shown that planets, following Kepler's laws, accelerated the same way that a rock falls to the surface of the Earth, then could not the force that caused a rock to fall be the same force that was causing the planets to fall toward the sun?

And couldn't that force therefore be produced by the matter toward which the objects were falling?

Haley undertook to ask the occult mathematician, Sir Isaac Newton, if such a proof were possible, and was told that he, Newton, had worked it out some time before.

Brahe's misperception that the moon sped up and slowed down in its orbit set in concrete the rather silly idea that the planets were falling toward the surface of the sun in the same way that rocks fall to the surface of the Earth. The fact that whatever the force was that makes objects fall is a straight line force, and the planets are going in near circles, didn't provide a pause for reflection. When things work so well in our minds, when the delusions we produce strike the soul of the consensus as being elegant, then looking up at reality as it exists becomes irrelevant.

Newton was able to state that he had already worked out the inverse square law with respect to falling planets, and according to Kepler's laws, the planets were dropping like rocks toward the sun.

BUT WHAT'S THE PROOF THAT THE FORCE IS PRODUCED BY MATTER?

Showing that the planets were accelerating the same way falling rocks accelerated on Earth did not show that the matter in the Earth and the sun was producing that force.

Could Newton mathematically demonstrate that the force was related to the matter toward which rocks and planets fell?

Newton reasoned that he could mathematically demonstrate that the force was related to the matter assumed to be producing it by showing that there was a relationship between the movement of matter and its size, even though Galileo's Leaning Tower of Piza experiment had proven there wasn't -- objects fall at the same rate regardless of weight. Newton, however, couldn't demonstrate the force on Earth because all he could do was drop rocks. Rocks didn't physically attract other rocks. They only moved toward the Earth.

Planets and moons were something else, though.

If Newton could show that the planets moved in accordance with a force that was proportional to the amount of matter in the planet and that force turned out to diminish in the same manner as the force that caused rocks to drop, inversely with the square of the distance over which it was acting, then the force, ***tied to matter by its proportionality to matter***, had to be a property of the matter.

The basic problem Newton faced was trying to figure out how much matter there was in a planet in order to compute the amount of force the planet produced.

NEWTON USES VOLUME TO DETERMINE THE AMOUNT OF MATTER IN A PLANET

Newton conceived of matter as made up of uniform balls, little units of matter that were evenly spaced throughout the matter (much as we think of magnets as made up of little magnets that are evenly spaced throughout magnetic matter). With matter made up of equally spaced units, if gravity were a property of matter, then the amount of gravity would be proportional to the amount of matter present.

By mathematically demonstrating proportionality, Newton would show the force to be a property of matter.

Newton chose the moon to prove that the force was proportional to and therefore a property of matter because, while the moon's orbit was only barely measurable, the orbits of the planets weren't measurable at all, Kepler's exacting measurements upon which his laws and Newton's enterprise was based being measurements of the mind rather than measurements of anything happening in physical reality.

Newton computed the amount of matter in the Earth and the moon by volume, equated that matter with the force that makes objects fall to the surface of the Earth, assumed the force extended out into space and therefore mathematically diminished it by the square of the distance between the Earth and the moon to obtain the force the two were exerting on each other, the force that was causing the moon to fall toward the Earth at an accelerating rate.

However, any idiot could see that the moon wasn't falling toward the Earth at an accelerating, or at any rate!

If the moon weren't falling toward the Earth, then there must be a force that was keeping it from falling to the surface of the Earth.

This force, the force that was keeping the moon from falling to the surface of the Earth, was the force that allowed Newton to construct an equation that would demonstrate that the force was proportional to, and therefore a property of matter.

NEWTON MAKES UP A FORCE TO PROVE THE FORCE HE MADE UP

Newton was attempting to prove that the force that caused a rock to drop extended out into space to affect the movement of the planets, affirming that they were falling toward the sun in the same way that rocks fell to the surface of the Earth.

If the moon was falling toward the Earth as a result of that force acting on it, then, because it wasn't in reality falling toward the Earth, there must be another force that was keeping it from doing so.

Reaching back into the history of nondisprovable ideas, Newton found Galileo's inertia, the force Galileo made up to prove the Earth was spinning, giving the appearance of the sun moving across the sky. Newton said that if an object can spin in space unless a force acts to stop that motion, then it can also travel in a straight line unless a force acts on it to change that motion.

Thus, the moon would be traveling in a straight line if it weren't in orbit around the Earth and the moon would fall to the surface of the Earth if it weren't traveling in a straight line. Thus, the force causing the moon to fall must be equal to the force making it move in a straight line.

This was where Galileo's flight of fancy about the perpetual motion of heavenly bodies caught up with Brahe's misperception that the moon could speed up or slow down in orbit.

The force making the moon move in a straight line was easily computable (as is anything that is nonexistent). All he had to do was take the moon's speed, multiply it by the amount of matter in the moon as determined by volume and he would have a force that equaled the force he had already computed that was making the moon fall toward the Earth.

If the two forces exactly matched, then the hypothesis that the force was proportional to matter would be proven because it would predict the moon to be falling toward the Earth in accordance with an inverse square law. Being proportional to matter, the force would have to be a property of the matter to which it was proportional.

The straight-line momentum of the moon would be exactly overcoming the force that made rocks drop to the surface of the Earth so that the moon was in a state of Galileo's perpetual motion, like one of his little balls on a frictionless path around the Earth, circling the Earth in frictionless space for almost forever.

Newton had to assume away a current cause for planetary motion in order to have something to measure, momentum, a planet moving in a straight line. If he didn't have anything to measure, then he couldn't measure what he couldn't measure to start with, the force that made a rock drop making the moon fall!

NEWTON FAILS

If Newton had been able to show the equivalence of the forces as determined by the amount of matter in the Earth and the moon it would have been pretty convincing evidence that the force was proportional to and therefore a property of matter. Not conclusive, by any stretch of the imagination. After all, the force could be a result of something the matter was doing, say vibrating at an atomic level.

But to be convincing at all, the equating of the two forces by the inverse square law had to be predicted from the amount of matter. Otherwise, there was no way to connect the force to the matter, or to say the force extended out into space to affect the motion of the moon or the planets.

But establishing a connection between the force and matter was moot because the equation simply didn't produce the intended result. The moon was too large to stay in orbit if the amount of gravity it produced were to be proportional to the matter that made it up by volume.

Newton first made Halley Astronomer Royal, and then hounded him to death (his own) trying to get measurements of the moon's motion that proved the hypothesis that the same inverse square law that controlled how objects fell on Earth also applied to keep the moon in its orbit.

NEWTON'S HEIRS REVERSE REALITY

In the century after Newton's death, the orbits of the planets become measurable, and while Kepler's Laws were never questioned, are in fact used to this day to plot the orbits of planets, applying Newton's failed proof that the inverse square law controlled the movement of the planets to the planets themselves demonstrated that it not only didn't apply to the planets, it didn't even come close.

The difference between how big the moon should be if it balanced the equations using the inverse square law, and how big it actually was, was not great enough to disbelieve the conclusion. Many explanations were given for the discrepancy between hypothesis and fact, some surviving to this day. Thus, some held the idea that the moon was made out of green cheese because cheese had a bunch of holes in it, allowing for the apparent matter that couldn't be there if Newton's proof were correct. Others assumed, because the backside of the moon never faced the Earth, that it was concave, accounting for the missing matter in that way. Still others contented themselves with the thought that the moon was filled with caves, fantasies that were carried into the earliest science fiction movies.

Some believe to this day that the Earth is hollow, accounting for the missing matter by turning inward, as it were.

However close the moon came, the planets didn't even come close to moving as predicted by Newton's hypothetically extended inverse square law. Nothing in the solar system worked out in accordance with the predictions of the Newtonian equations that assumed that gravity was proportional to and a property of matter based on volume and extended out into space to affect the motions of the planets, decreasing inversely with the square of its distance as it did so.

What to do?

Well, the encomiums for Newton had already been carved in stone. There was no question that he was right. So either the universe was wrong, highly unlikely, or Newton had overlooked something.

Think, think, think. What could Newton have overlooked?

Ah, it finally dawns on his heirs. Not only is matter not distributed evenly throughout a planet, nothing is writ in heaven that all planets be made up of the same matter.

Newton had overlooked the hard, earnest fact in reality of density!

But, Newton knew that factoring in density would prevent him from computing the amount of force a planet produced, and not knowing the amount of force a planet produced meant that he couldn't show that the planets moved on the basis of an inverse square law and that the force was therefore a property of matter.

If you don't assume volume, you can't show that there's a connection between a falling object and the matter toward which the object falls by showing the planets move as a result of an inverse square law predicted by the amount of matter.

Oh, well, said the eighteenth century sages, we can't have everything. We know that objects accelerate in accordance with the inverse square law when they fall on the surface of the Earth, that's directly measurable. And we know that Kepler's laws dictate that planets accelerate on the basis of an inverse square law. That's the law.

So what's to prove?

The value of Newton's equation is not that it proves anything, his heirs concluded, but rather that it provides a means to compute the amount of matter that makes up a planet!

Newton's heirs, not being able to balance his equations with matter predicting the hypothetical force, the inverse square law, decided to use the hypothetical force, the inverse square law, to predict nonverifiable matter, what's in the center of planets.

This reversal of logic, a mirror misanalogy, starting out using the amount of matter assumed to be in a planet to prove that an inverse square law controlled a planet's movement, that the planets were falling toward the sun as a result of the same force that made objects fall on Earth and ending up applying the inverse square law to the orbits of planets to determine the amount of matter in the planets is absurd to say the least, leading it to qualify as one of the greatest scientific misconceptions of all time.

NONDISPROVABLE DELUSIONS RULE

When we apply the inverse square law to the movement of planets and stars, we end up eliminating matter, or worse, creating it. Thus, Jupiter becomes a ball of gas, comets, belching fire every inch of the way, showing up on photographs as solid chunks of red hot rock, become snowballs, the sun, the matter that is supposed to be attracting all of the planets, becomes helium, lighter than air.

There is no way to determine what exists at the center of the planets or in the sun, so these become eminently fertile areas on which genius can pontificate without any chance of a Copernicus coming along and embarrassing them by proving them wrong. Computing the assumed motions for all of the assumed stars in a galaxy creates ninety percent more matter in the galaxy than can be seen, a safe assertion that sets off billion dollar searches for this "dark matter".

And if making absolute determinations about what is in the center of planets seems too dangerous for all but the most timid, we can just apply the inverse square law to space and come up with the ultimate in safe assertions, black holes from which no light can escape. If no light can escape from nothing, then no one will be able to find nothing, and if you can't find nothing, you can't disprove nothing, creating Empirical Science's answer to elves.

The matter created and destroyed by measuring motion isn't just ordinary matter, it is the matter that produces gravity, so it has to have a more exalted designation than just matter, else someone might figure out it was basically meaningless.

Besides, the term matter is ambiguous because it doesn't take into consideration density. When we look at something, we have no way of knowing what's inside the something, how much of whatever is there. On Earth, we can find out specific gravity by having the matter displace water, but we can't expect empiricists whose job it is to accurately measure what they see in the universe and report on the results to dump planets into oceans. They wouldn't fit.

So mass is the density that results from applying the inverse square law to a planet's motion.

This results in mass/gravity, a meaningless phrase with no basis in reality that we use to answer the question, what makes objects fall.

HOW TO DEAL WITH THE DELUSION

Thoughtful people have treated this situation a number of ways. Some use electrostatic surface tension to trick themselves into believing in the reality of mass/gravity. There is evidence that Newton was among these.

Some have claimed capillary action in vegetation as proof of mass/gravity, presumably because it only works in plants when the massive sun is overhead.

Others have said, well, mass/gravity may not be the answer, but it is as close as we are ever going to get, so let's let it go at that.

Still others have out-occulted Newton's occult concept where the force acts without exerting work, causes real movement in reality without depleting itself, doesn't exist between its source and its effect, and exists an infinite distance from that source. They come up with multidimensional concepts, quantify that which doesn't objectively exist, space and time, to produce inconceivable warps in nothing illustrated with Galileo's balls rolling around on checkered rubber sheets.

Many in the profession cry, "Abandon Newtonian Mechanics? Not on your life. It's just too useful. How else would we be able to tell you what's in the center of the Earth, the moon, the sun, even in the center of shrouded planets like Venus and Jupiter!"

And finally many say, who cares? After all, objects drop, planets move, the sun comes up. What difference does the answer make so long as it doesn't raise other questions that can't be answered? In fact, why not just leave the answer alone so that other questions won't be raised? Knowing that the Earth goes around the sun hasn't added anything to the progress of mankind. The sun's still there and we're still here, and it's only recently that the knowledge has been useful in sending rockets to the planets.

WHY SHOULD WE CARE WHY OBJECTS FALL?

When we think we know something, then it becomes invisible.

If we think we know what makes planets move, then we will never attempt to uncover the force that is actually making them move.

If we think we know what makes a rock fall, we will never attempt to find out what the force is that actually makes objects fall.

Technology involves testing reality to see if it is amenable to being shaped to our purposes. Science, its methodology, whatever that is, the empirical process, the use of mathematics to dictate what we see, has hidden behind technology, pointing to technology as its fruit when, in fact, scientific conclusions have more to do with the absence of technology than the existence of technology. Scientists are extremely reluctant to test reality because of the embarrassment of being proven wrong, and are more comfortable in realms where no testing is possible, determining what is at the center of planets, in deep, dark space, or declared unmeasurable in atomic structures. It has only been during the last half of the twentieth century when technology has outstripped scientific theory that we have seen advances transcending the negative dictates of the nebulous nontheories dealing with magnetics, electricity, heat and light.

But gravity remains out of reach because we all know what it is, a property of matter, and we all know that properties simply exist, gold is yellow, steel is hard and ice is cold, and that's the scheme of things and nothing can be done about it.

We use technology to extend our range of survivability in reality, using our minds to bend reality to our needs, evolve it to our purposes rather than risk going missing, becoming extinct, as a result of changed reality. If we think we know something when we don't, we can't test reality, and thus we can't bend reality to our needs.

If we think we know reality, and we don't, we risk extinction

IT'S NOT THE ANSWER, STUPID, IT'S THE QUESTION

The most obvious fact in reality is the force that holds us to the surface of the planet. Not knowing what that force is, how it originates, how it moves over distance, how it acts on matter, is absurd, reducing us to savages playing with the toys we accidentally create with our minds.

There is a hypothesis in physiology called Bernstein's Hypothesis. When electrical impulses were probed as a source of bodily action in the nineteenth century, it was thought that electricity traveled on the outside of a conductor. As nerves weren't like conductors, and no current could be measured on their outside, an idea was cooked up that made nerve impulses travel similar to light. A state of polarity was assumed in the nerve which was disrupted by sensory stimuli. Nothing moved down the nerve, rather the disturbance in polarity traveled down the nerve.

Today, when we know much more about electricity, at least to the extent that it moves in conductors rather than on the outside of conductors and that conductors need not be limited to copper or silver, we still hold the view that nerve impulses don't exist in and of themselves. This results in extraordinary conceptual blindness when it comes to how the physical mechanisms of the body work, leading to universal disclaimers on product information statements dealing with the clinical pharmacology of drugs that the exact mechanics of the effect the drug produces have not been definitely demonstrated.

We think that dropping an object demonstrates all that is needed to be known about the effect, that gravity is the answer to the question why objects fall, when it is the mechanics of the force we need to know about, not its effect.

What gravity is, how it is produced, how it moves and how it acts on matter to move it, these are the questions we need to be asking.

There is nothing wrong with attempting to come up with answers based on the information available, but the information available to Galileo, Brahe, Kepler and Newton was next to no information at all.

There is something wrong with being too lazy to attempt to come up with answers based on current information.

Our slavish belief in the unsupportable conclusions of dead men who knew nothing is criminal because it will lead to our own extinction.

2. THE OCCULT NOTION OF MOTION

OUR HALF-VAST KNOWLEDGE

We are told frequently that there isn't much else to know about physical reality. Of course, there are some details about how fast the universe is expanding, and some disagreement about how old it is, and there seems to be a bit of consternation whether a space traveler will cease to age in proportion to the speed of his spaceship, and, of course, there are a few problems with whether time can move backwards as well as forwards, and doubt about whether the mind, which has no connection to the body because it doesn't exist, can create alternate universes with its every thought.

Other than these minor problems, all that we need is one or two more terms to neatly wrap up the explanation for all physical reality and existence into one equation that will fit on everybody's forehead so we will be able to walk around and see eternity on everyone's face.

There is one little hitch with all this godly knowledge our genius has discovered.

We don't have the foggiest idea why anything in the universe moves!

We have no explanation for why objects fall when the are dropped, why the Earth rotates on its axis, why a planet orbits the sun, how the suns configure themselves into galaxies, we don't have any explanation for the movement of the flames that make up the fires that heat our food and warm our bones, we are clueless as to what causes the wind to blow, we are mystified as to the cause of the ocean currents and how light gets from one place to another, we have no idea at all what electricity is or how it moves, what causes two magnets to move toward or away from each other, what gets radio and television signals from their source to our sets, heck we don't even know what keeps our lungs operating or our hearts pumping.

We can ignore all of these little ignorances because our attention is focused on producing telescopes that will, so we believe, extend our visual range out to the edge of the big bubble that is the universe so that we can actually witness its beginning, the creation of space and time and the formation of matter out of a massive explosion that would obliterate it on Earth. When we aren't exploring that which is too big to measure, we are defining that which is too small to measure, manufacturing atomic particles, subatomic particles, particles that make up subatomic particles, and particles that make up particles that make up subatomic particles.

Why should we worry about little things like what makes the sun come up, or what makes objects drop when we can focus our attention on the nature of event horizons in black holes or the charm of quarks?

WE DON'T LOOK FOR ANSWERS BECAUSE WE THINK WE HAVE THE ANSWERS

The reason that we drop the ball explaining the obvious realities of our physical existence, why we can ignore the fact that we don't know the source of the obvious motions in reality, is that we have been fed occult notions for those motions and accepted those occult notions as rote fact, instantly reciting dogma whenever questions of motion arise.

"What is the source of the force that makes objects fall?"

"Gravity."

"What is gravity?"

"If you don't know what gravity is, let's take a hike up the Empire State Building and I'll use you to demonstrate it!"

Everyone knows what gravity is, it's what makes objects drop. Anyone that doesn't know this is obviously an idiot. Idiots are surplus matter and can be disposed of by using them to demonstrate gravity.

Of course, the idiots are the ones that believe that naming the force that causes dropped objects to move to the surface of the Earth produces understanding about the nature of the force that causes those objects to move.

Such an idiot will respond to the question, "What is gravity?" by the rote, "Gravity is a property of mass."

This displays a considerable depth of knowledge, akin to being able to recognize that it's better to be hit by a pillow than by a baseball bat.

If you ask, "What is mass?" then you find out from the idiot that mass is what produces gravity.

Nothing about any of this tells us anything about the nature of the force that is causing a dropped object to move, but because the responses appear to be answers, after all, no one is saying the answer is wrong, they are just as good as answers.

When we know the answer to something, then we don't have to ask the question!

What makes the sun rise gets a little stickier because it requires a two level explanation. For years we thought the sun did just that, rise in the east, move across the sky and then set in the west. One of the biggest mistakes in our perception of physical reality is considered to be the acceptance at face value that because the sun moved across the sky, it actually did move across the sky. In fact, when we thought the sun moved across the sky, we didn't much care what caused dropped objects to fall because where else would they fall but around what everything moved, the center of the universe.

It was only when we found out that it was not the sun that was moving across the sky, but rather the Earth that was spinning on its axis giving the sun the appearance of moving across the sky that we began to worry about what made dropped objects fall, because the only thing that was keeping us from being hurtled off the surface of the rotating Earth into the endless darkness of space was the force that was causing dropped objects to fall.

But what is the source of the force that causes the Earth to rotate on its axis?

IF WE CAN'T SEE THE FORCE, THEN IT ISN'T THERE

Galileo, the foremost proponent of Copernicus, the person credited with overturning the misconception about the sun's motion, set out to explain what the force was that was causing the Earth to rotate.

But there was no apparent force causing the Earth to rotate!

In Galileo's time, this wasn't too much of a problem because it was universally believed that the Creator had produced the matter in the universe with its motion installed. Why create matter without motion when both tasks could be accomplished with the same act?

Thus, the fact that the Earth was spinning on its axis wasn't so much a problem for Galileo. His real problem was why it was still spinning.

This change in perspective is important enough to give us pause because it is at the basis of all of our misconceptions about motion today. It is similar to the question of whether the sun was or was not moving because how we perceive the cause of motion controls our perspective of the universe.

The person that was making the rules, so to speak, the man who was setting the groundwork for all subsequent understanding of motion in the universe was not concerned with what was causing that motion because he believed the cause to be obvious, God caused the motion, he was merely concerned with why the motion hadn't stopped.

The difference in perspective is the difference between knowledge and ignorance. We look at something moving and ask, what is making it move, or we look at something moving and say, what is keeping it from stopping!

Choosing ignorance, asking what is keeping something that is moving from stopping, produces two major obstacles to understanding physical reality.

The first obstacle is that God as the cause of motion puts the cause of motion in the past. We can't see the cause of the motion, and we certainly can't see God, He's not a palpable presence.

By producing a nonpresent source for the force, a historical source, we have placed the cause of motion outside our direct control, outside our ability to manipulate it. We cannot analyze what no longer exists, the creation of the motion at sometime in the past, and we certainly cannot directly affect that source if it is historic, if it no longer exists.

Worse, by creating a historical force beyond analysis, we can make up any old stupidity we want to explain the force. If God becomes unfashionable, we can come up with some other explanation. If it is not a current force, if it no longer exists, who cares? We can't do anything about it anyway.

The second obstacle to understanding physical reality that looking at motion as something that occurs unless it is opposed is that instead of looking for a cause that is keeping something moving, a current cause of the motion, we assume away the cause of the motion and look for a reason why it hasn't stopped moving.

What we have done is assume away a current cause of the force because we cannot visibly see what is causing the movement. *Instead of looking for something that exists, the cause of current motion, we spend our time looking for something that doesn't exist.*

GALILEO, THINKING HE HAS DEMONSTRATED INERTIA AS THE FORCE CAUSING ROTATION, CREATES THE OCCULT NOTION THAT HISTORIC FORCES PRODUCE CURRENT MOTION BY OCCUPYING THE MATTER DOING THE MOVING

With the cause of motion clear, God put the Earth in motion when He created it, Galileo only needed to demonstrate that the current motion was a result of this historic cause. He merely had to prove that the historic force was still causing the current motion.

How could he do this?

By showing that there was nothing to interfere with the historic motion, nothing to cause that historic force to stop acting. In short, all he had to do was prove a negative.

How do you prove a negative?

Phrase it in the positive!

The Earth, set in motion by God at the creation of the universe, will continue to rotate on its axis unless a force causes it to stop rotating.

What could stop the Earth from rotating on its axis?

Friction.

Because empty space is frictionless, it is after all, empty, and because the Earth is rotating in empty space, it is rotating without friction and thus the motion imparted to it by God continues to cause it to rotate.

Because Galileo was phrasing his proof in terms of what would keep the Earth from rotating, he called the force that maintained the Earth's rotation inertia, a resistance to a change in historic motion placed in the Earth by God.

How could this fact of inertia be proven?

A half a century earlier, Tartaglia had found that the optimum angle for distance to fire a cannonball was forty-five degrees. A fired cannonball had its parabolic trajectory cut off by hitting the ground. Galileo, rolling balls down a plane inclined at forty-five degrees, found that when the balls rolled off his workbench, they described Tartaglia's parabolic trajectory and concluded that without the Earth to cut off the forward motion of the balls, the forward motion would continue forever.

Because the forward motion would continue forever, Galileo reasoned that if he could construct a frictionless path that circled the Earth, then the force imparted to the balls by the inclined plane would cause the balls to roll around the Earth forever.

Inertia was, therefore, not a concept but an actual fact. A frictionless pendulum would swing forever, a spinning coin with a frictionless contact would spin forever, a ball in a frictionless circular path would roll around the Earth forever.

The only fly in Galileo's ointment of frictionless rotation in empty space as the basis for saying that a historic force was causing the Earth's rotation was his attempt to demonstrate that the tides resulted from the Earth's rotation. Even though the tides do, in fact, operate as if the Earth's rotation were driving them, all that sloshing back and forth would create friction which would cause the historic force causing the Earth's rotation to disappear as fast as love letters in the sand.

Galileo retreated as quietly as possible.

WHAT'S WITH THIS FRICTIONLESS BULLCRAP?

It's very difficult today for us to visualize why the Earth couldn't rotate forever in frictionless space without a current force. That's the second level in understanding why the sun doesn't move around the Earth.

"What makes the sun move?" little Joe asks.

"It doesn't move. It just appears to move because the Earth is rotating on its axis."

Little Joe gives this some thought. "Well, what's making the Earth rotate on its axis?" he finally asks.

"Hmmm. That's a complicated questions. Better ask the internet."

"What causes the Earth's rotation?" little Joe types into the appropriate block of the Scienceanswer.org website.

"Angular momentum." the Gurus respond.

"What's angular momentum?" little Joe types after an appropriate pause.

"It's, um, like, when an ice skater is spinning, and she draws her arms in and spins faster. Angular momentum is conserved in closed systems."

Little Joe pulls up Excite and types in "Live Nude Girls" while the Gurus at Scienceanswer.org go back to their discussion about how they can promote more respect for the scientific process among the young.

The reason that we believe that something can move forever in frictionless space as a result of historic forces is that it is intellectually true.

There is no frictionless space, and there is no historic force, but if there were a historic force and it put something in motion in frictionless space, then that something would most certainly move forever.

We visualize the world with our recall because it is with our recall that we form the concepts that provide meaning to the facts that make up the world. Thus, even though there are no facts in reality that would provide a basis for our recall that the Earth rotates because it received a force in the past and that force has continued to act because there is no other force opposing it, we believe it.

If we didn't believe it, then we would have to go out and figure out what the current force was that was causing the Earth to rotate. That might require some mental effort, and, because everybody has the same recall, everyone believes the Earth rotates as a result of an unopposed historic force, produce personal danger as well. Making assertions that oppose people's beliefs tends to get one strung up on the nearest yardarm.

Of course, another way to get yourself strung up is to claim that you have invented a perpetual motion machine, because such a machine, for openers, would require frictionless movement which can't in reality exist, but that just emphasizes how unimportant reality is in the face of our great ideas.

Tartaglia's cannonballs did not fall to the surface of the Earth because of the friction in the air, they fell to the surface of the Earth because the Earth in some way produces a force that causes objects to fall to its surface. Galileo's balls did not describe an arc that would produce forward motion forever, they hit the damn floor and stopped moving!

The surface of the Earth causes objects to come to rest with respect to the forces acting upon them, and the primary force acting upon them is what is called gravity.

For Galileo to make the mathematical claim that a frictionless pendulum would swing forever, or a spinning coin would spin forever if its contact were frictionless, or that balls would roll around the Earth in a frictionless path forever ignored the effect of gravity on the frictionless motion.

Even if frictionless motion existed, the force of gravity would cause the pendulum to stop swinging, the coin to stop spinning and the balls to stop rolling.

THE LEGACY OF THE BELIEF IN FRICTIONLESS MOTION

By assuming away the cause of motion as historical, we contradict the only measurable reality we have, the way motion is measured on Earth.

On Earth it takes force to get something in motion because gravity is omnipresent. Once something is in motion, then it comes to rest with respect to the forces acting upon it. A pendulum, set to swinging, will slowly stop swinging as it comes to rest with respect to gravity. A cannonball bolts out the end of the cannon and immediately starts to come to rest with respect to the forces acting upon it.

Nothing on the face of the Earth moves without the application of force, and nothing on the face of the Earth continues to move without the application of force.

Objects come to rest with respect to the forces acting upon them!

However, Galileo didn't look at it this way. He said something continues in motion unless friction acts to retard that motion. He said a force sets something in motion, and then it becomes, in effect, a part of the matter, a property of the matter. The cannon or the inclined plane imparts an actual force that becomes a part of the matter doing the moving so that when the initial force is removed, there is still a current force moving the matter.

Instead of saying what is actually happening, that an object is at rest with respect to the current forces acting upon it, a current force has been applied to cause an object to move with respect to the current forces acting upon it, and when the applied force has been removed, the object will continue to move until it returns to rest with respect to the current forces acting upon it, Galileo said the matter obtains inertia which causes it to continue to move until another force acts on it.

He made up a current force, inertia, to explain the movement of matter that was merely coming to rest with respect to current forces.

As a result, instead of matter moving in response to applied force in space as it does on Earth, matter moves in response to the failure to apply force and only changes that movement when force is applied to change that movement.

We had two choices to make when we looked at the moon moving across the sky, and when we realized that the Earth was rotating on its axis and the Earth and the planets were thus orbiting the sun.

We could say, there is a force that is causing all this movement and our job is to sit down and put on our thinking caps and come up with the cause of the force, a physical description of what it is, how it is generated, how it moves through space and how it acts on the matter to cause it to move. We could say, this is incredible stuff, all this movement. It is a priority to figure out what it's all about because surely this knowledge is central to our understanding of the universe and thus our continued existence in that universe.

Or we could say, "Well, look there at all that movement, duh, there's nothing there so therefore there's nothing there causing the movement. If there's nothing there causing the movement, then all the movement must have been caused in the past, God, swirling masses of gas, big bangs, it doesn't really make much difference because it can't be seen so it isn't there anyway. If anyone questions us, well, no one has ever proven there isn't a God, and no one is going to prove there wasn't a swirling mass of gas or a big bang or something to set everything in motion."

We had the choice of using our minds and coming up with some answers that actually explain the movement we measured, or making up a bunch of dumb cluck nonanswers that don't explain anything but which keep the questions from coming up.

"Where did the universe come from?"

"God created the universe."

"Didn't you hear, God is dead?"

"Well, then it condensed out of a swirling mass of gas."

"But only the solar system is swirling."

"Then it came from a big bang."

We had the choice to accept that matter in space works like matter on Earth, it requires current force to produce movement, and then to look for that current force, or we could accept nonexplanations for the obvious realities in our existence.

We chose nonexplanations!

HOW THE ARGUMENT FOR CURRENT FORCE WAS LOST

Four hundred years ago when these decisions were being made for us by dead men who knew nothing, there actually were strong proponents for a current force at work moving the planets.

Kepler, as a result of his mentor Brahe's mistaken measurement that the moon sped up in the summer and slowed down in the winter, believed, as a result of applying Brahe's mistake to the orbit of Mars that the planets sped up and slowed down as they moved around the sun.

Now, if a planet speeds up and slows down, what conclusions have to be made about Galileo's historical force imparted at the creation of the universe? The fact that Galileo had apoplexy at Brahe's measurements of the moon speeding up and slowing down should provide a hint.

If something is speeding up and slowing down, then it cannot be operating as a result of a historical force, it has to be operating as a result of a current force, otherwise there would be no way for the planet, or the moon, to regain the force that it had lost when it slowed down.

In the face of Galileo's total denial on the subject, Kepler argued that because planets had to speed back up after they slowed down, there was a current force causing movement in the solar system. He felt that the force that produced the current motion in the solar system somehow resulted from the sun's emissions.

Besides being a religiously risky assertion, after all, a current force denied that God had created motion when he created the matter in the universe, Kepler had no way to respond to the observation that if the sun's emissions were the source of motion for the planets in the solar system then where was the source of the moon's motion around the Earth.

In the mid-twentieth century, astronomers were still claiming the negative that there was no evidence that planets were internally hot and thus produced emissions. In fact, today it is generally believed that thermal equilibrium prevails, that the Earth only gives off the heat it receives, having maintained its temperature in empty, temperatureless space for some five billion years, so it is easy to see where Kepler's idea that it was something about the sun that caused the planets to move on a current basis ended up.

It was summarily dismissed!

This is one of two perfect examples that point out the fallacy of creating laws based on limited information and then applying the laws to all subsequent facts. There is no argument today that the Earth is hot. It is only the rare idiot that gets up and claims that volcanic magma results from the friction of plates beneath the surface of the Earth. The Earth's core is hot as hell, and the Earth is an emitter, clearly removing objections to Kepler's claim of a current force based on emissions, but uselessly so, since historic force, objects travel in straight lines unless, is a law, and no current force is necessary. The other example, of course, is making up the laws that apply to magnetism, heat, light and electricity before having any knowledge of induction, and then continuing to apply those made up laws in spite of induction.

Descartes avoided the problem with the moon moving around a nonemitting Earth by claiming that the movement of bodies in space was circular, that circular movement in water caused vortices, and thus the planets and moons were moving in the vortices created by their own movement. Leibnitz later reflexively subscribed to this theory in opposition to the historical force Newton adopted from Galileo.

The problem with planets and moons moving in the vortices created by their own movement was a chicken and egg problem. Which came first, the movement of the planets to cause the vortices or the movement of the vortices to cause the movement of the planets.

There being no way to argue that current force produced current motion in the solar system, the door was open to Newton's occult notion that momentum and gravity could affect each other without being used up.

OVERCOMING THE CHICKEN AND EGG PROBLEM BY INCORPORATING THEM IN A HISTORICAL FORCE

Huygens, the guy responsible for the water wave theory of light and thus much of our total ignorance about the nature of physical reality, stunned the scientific world by inventing the pendulum clock. By putting weights on a few pulleys and chains and regulating the effect of the weight on the pulleys by gears which also could be connected to hands that measured the timing of the whole process, he produced a clock that worked forever, the only requirement being that a servant come along periodically and reverse the effect of the weights, pulling them back up so that the Earth's attractive force could keep the whole thing operating.

The weights of a pendulum clock descend so slowing as a result of gravity's gear-muted tug that their movement is hardly noticeable behind the rhythmic swinging back and forth of the ever-present pendulum.

If we ruminate about the great thinkers in history while watching the pendulum swinging back and forth, we could say, after the fashion of Galileo, that the pendulum, without friction, would just swing back and forth forever.

These daydreams, nonreligious in nature because they are based totally on observation, might even be extended to a bouncing ball. If the friction of the atmosphere, and the friction between the ball and the floor were removed, wouldn't the ball bounce forever?

After all, what was there to slow it down?

Well, there was the same thing that slowed Tartaglia's cannonballs down. As soon as the cannonballs were shot out of the end of the cannon, they began to fall to the Earth because of the force of gravity.

But what if the surface of the Earth weren't there? What if they were, like the balls rolling off Galileo's workbench, describing parabolic courses which would retain their forward motion forever? Wouldn't shooting a cannonball off into space result in the cannonball being released from the friction of the atmosphere and also from the effect of the Earth's gravity?

Like the frictionless pendulum that would swing forever, wouldn't the cannonball, if given just the right momentum, continue forever around the Earth just like the moon, and just like the planets move around the sun?

Sure it would, and that's just what is happening. The moon and the planets received their motion in the past and because they are moving in frictionless space, they just keep right on moving like the pendulum would move if the pendulum could move without friction.

Huygens' swinging pendulum, Galileo's frictionless balls, Tartaglia's cannonballs, all these things were eminently measurable quantities and provided a basis for the measurable conclusion that the objects in the solar system were moving in frictionless space as a result of a historic force.

Because the measurements hid the fact that the cause of the movement remained unexplained, the explanation was occult. The mathematical bullshit hid from view the continued fact that the source and nature of the force that produced motion in the solar system remained unknown.

The cannon charge was the source of motion for Tartaglia's cannonballs, the inclined plane for Galileo's balls, the weight for Huygens swinging pendulum.

And God was the source of the force for Newton's planets!

But what about the fact that the planets sped up and slowed down? After all, Newton had used the inverse square nature of the measurement produced by the fact that Kepler's planets were supposed to describe equal areas in equal times for his explanation that the planets and moons were falling toward the sun and the planets like rocks fall to the surface of the Earth, inversely with the square of the distance over which they fall.

Newton had to concede that they were speeding up and slowing down.

How could they be speeding up and slowing down as a result of a historical force?

Not to worry!

As the planets fell toward the sun, they were speeding up weren't they? Their speed increased with the square of the distance over which they traveled. Kepler's laws said so even though Brahe's measurement of the moon and Kepler's measurement of Mars had them slowing down in the winter and speeding up in the summer, *which would mean that the planets' speed increased as they fell away from the sun and decreased as they fell toward the sun, the opposite of Newton's fantasy.*

In spite of this, the result of Newton's idea was the belief that the planets regained their speed by falling toward the sun. As they moved away from the sun, just like a bouncing ball that is moving up away from the Earth, the planets lost their speed, slowed down, but then, just like a bouncing ball, when they got as high as they could go, they started to fall back toward the sun, and regained their speed.

The chicken and the egg were incorporated into opposite sides of an orbit's measurement, force in on the fall, force out on the rise, and in the process the lack of knowledge about the current cause of the force was totally obscured.

But a bouncing ball eventually comes to rest on the surface of the Earth. Why haven't the planets come to rest on the surface of the sun?

GRAVITY AND MOMENTUM ARE NOT USED UP WHEN INTERACTING THEREBY MAKING THEM MORE OCCULT THAN GOD

A freely swinging pendulum eventually comes to rest as a result of the Earth's force of attraction, gravity, acting upon it. A spinning coin is eventually upended by gravity slowing its spin. Galileo's rolling balls eventually come to rest with respect to gravity as do Tartaglia's cannonballs.

Objects come to rest with respect to the current forces acting upon them and gravity is the most effective force in the universe for bringing objects to rest. Gravity causes objects to stick motionless to whatever surface is between the source of the gravity and the object and it takes a great deal of force to unstick those objects.

But that is only in the real world, the world we measure here on Earth.

In frictionless space, even though frictionless space doesn't exist anywhere in the universe, things work differently.

When it is not doing anything, when it is merely diminishing, gravity is used up.

However, when it is causing the planets to accelerate, or when it is slowing the planets down, causing them to decelerate, it is not used up. The gravity causing the Earth to accelerate toward the sun does not prevent the same gravity from causing Venus to decelerate.

So we are dealing with a force that is not used up when it does something but is used up when it doesn't do anything.

That's clearly beyond the realm of human understanding, about as occult as you can get.

But what's more occult is the momentum the occult force of gravity is acting upon.

The planets are traveling in a straight line. That's clear from Newton's laws of motion. All matter travels in a straight line. It was the motion the matter received when it was created by God, swirling masses of gas, big bangs, whatever.

The only thing that could keep a planet from moving in a straight line was a force acting upon it to keep it from moving in a straight line.

Gravity is the only force in the universe, it was created as a property of matter by God when God created the universe, so gravity is the reason that, although everything in the universe is going in a straight line, it isn't. Gravity bends the straight-line motion of the planets and the moons into circular motion.

But because gravity is not used up when it does so, no energy is expended in bending the straight-line motion into circular motion. With none of the gravity's energy expended in bending the straight-line motion, none of the momentum produced by the straight-line motion is used up by being bent into circular motion.

Thus, unlike the bouncing ball which eventually comes to rest with respect to the forces acting upon it, when the planets and moons decelerate as they move away from the sun and the planets, they are not losing any energy doing so.

As a result, unlike the bouncing ball, they can regain all the height they lost.

The momentum of the planets and the moons, not being used up when it exerts itself against the force of gravity, is also beyond the realm of human understanding and therefore it is just as occult as the gravity it is not using up.

But that's not as occult as it gets. The occult notion of motion pushes beyond the realm of human understanding even further, creating matter where none exists and eliminating it where it does!

THE RESULT OF APPLYING THE INVERSE SQUARE MEASUREMENT OF FALLING OBJECTS TO THE OCCULT FORCES OF GRAVITY AND MOMENTUM

Newton attempted to show that gravity was a property of matter by mathematically demonstrating that it was proportional to matter. He set out to show that the momentum of the moon was equal to the force of gravity produced by the matter in the Earth and the moon. To do this he had to know the amount of matter in the Earth and the moon. By showing how much force there was in the Earth and the moon, he could show that that force, diminishing inversely with the square of the distance over which it acted, equaled the momentum produced by the amount of matter in the moon and thus, operating on the basis of the same inverse square law that caused objects to fall to the surface of the Earth proved, at least to the indiscriminate, that it was a property of that matter.

The whole enterprise depended on knowing the amount of matter in the Earth and the moon, a definite unknowable unless the assumption was made that the matter in the Earth and the moon was spread uniformly throughout.

Newton therefore made up his corpuscles of matter which he claimed were uniformly distributed throughout matter.

The whole business failed of proof, the moon having too much matter by volume to stay in orbit. It failed even more miserably when applied to the motion of the planets.

Thus, gravity was never proven to be proportional to and therefore a property of matter and there was no relationship demonstrated between the inverse square measurement of falling objects and the movement of the planets.

But what the hey, when we are dealing with historical forces there is nothing currently existing that can contradict anything we want to say. We can make up bullcrap to our heart's content, and who can gainsay it. We are, after all, dealing with the occult, the occult by definition is hidden from view beyond the realm of human understanding, so we can feel free to roam whither our fantasies carry us.

While it's true that no one can measure what's inside the planets and the moons, if we call the inverse square measurement of falling objects a law and apply it to the movement of the planets, then we can compute the amount of matter that makes them up. All we have to do is measure the circular motion and then compute how much matter would be required to cause that circular motion if the matter produced a force of attraction that diminished inversely with the square of the distance over which it acted.

If we can't use matter to prove that a proposed inverse square law demonstrates that the attractive force is proportional, and therefore probably a property of matter (as opposed to something the matter is doing, vibrating, oscillating, sweating), we can create an inverse square law to show how much matter is causing the gravity.

Why prove the inverse square law when we know that's how objects fall when they are dropped, and why prove gravity when we know that dropped objects fall.

We'll just assume that both apply to the movement of matter in space and use the result to compute the amount of matter causing the force.

Thus, measuring the orbit of the planets with the belief that those orbits are the result of the inverse square measurement of falling objects on Earth, the inner planets become planets with cores made out of nickel and iron and the outer planets become planets with cores made out of gas. In the meantime, the moon has a concave backside, is filled with caves, or is made up of cheese, and the crowning glory of Modern Science, fiery comets in the sky are, because the application of the inverse square law to their orbits dictates it, actually snowballs!

And, when we really get carried away by our fantasies, we can look deep into space and find that, without enough visible matter present to hold the stars into galaxies, those galaxies are actually made up ninety percent of dark matter, stuff we can't see and never will, and that, in addition, the fabric of space, the definition of which is blank, empty, frictionless, *nothing*, is rent throughout with holes which, because they are black, cannot be seen.

Hidden from view?

Beyond the realm of understanding?

This occult bullcrap makes God's presence palpable!

MALICE OR STUPIDITY?

It's one thing to look up at the sun crossing the sky and argue about whether the sun is going around the Earth or whether the Earth is going around the sun. For one thing, as far as appearances are concerned, it actually looks like the sun is going around the Earth. The millennia through which the subject never even got discussed attests to the power of this appearance.

And besides, what difference did it make? The relationship didn't become important until the mid-twentieth century when we started to shoot rockets into space.

Today, the question of whether there is a current force operating the solar system is not even a question. The solar system was put in motion by a swirling mass of gas and has operated unchanged for some five billion years, and will continue to operate unchanged for another five or ten billion years unless an asteroid (as opposed to a giant snowball like a comet) comes along and causes a temporary glitch which will wipe out life but leave the planet relatively unscathed.

Does it make any difference what operates the solar system, God, historical swirling masses of gas, or some little invisible guy with a crank and some very thin gears?

What's wrong with occult notions of motion?

Does not knowing the cause of motion make any difference?

It does make a difference!

The force that operates the solar system is the same force that makes objects fall to the surface of the Earth. If we view the nature of the force as occult, hidden from view beyond the realm of human understanding, then we will never obtain an accurate picture of physical reality, and without an accurate picture of physical reality, we will never be able to do what we evolved to do, move safely through that physical reality.

Consigning one source of motion, rotation and orbiting, to a historical force and the other source of motion, falling objects, to the status of a static property of matter, and then dismissing all movement as the interaction of the two, that orbiting objects would be going in a straight line but for the fact that they are falling and they would be falling but for the fact that they are going in a straight line, *when they are doing neither,* destroys the mind because it itself is a mindless assertion with no basis in reality.

We can't compare ourselves with the preCopernicans who walked around with a mistaken fact of a relationship of the movement of the Earth to the sun because at least they were mistaken about a fact.

We can't even claim to have made a mistake about a fact, what is causing the motion of the matter in the solar system, rotating and orbiting planets and falling objects, because we have simply assumed the cause away, said it doesn't exist, there is no current force other than that wrapped up in matter as a property like texture and color and hardness.

There is a thin line between malice, the intentional infliction of harm, and stupidity, the oafish negligence to duty that can be dismissed as relatively blameless. Being dumb and ignorant about what causes movement in physical reality is simple stupidity.

But claiming knowledge when none exists starts to border on malice.

The line to malice is clearly crossed, however, when the subject is totally ignored while mulcting billions out of the very people in whose name the knowledge is being sought for bogus explorations into matters such as black holes, event horizons, gadzillions of elementary particles, forces at the core of nature and formulae that explain everything.

What is the harm inflicted by the malice?

Knowing the cause of motion is at the basis of any attempt to come up with a consistent picture of physical reality.

Having a consistent picture of physical reality is an absolute prerequisite to our survival because our survival depends on our ability to manipulate physical reality to our benefit. Without an accurate picture of what we are manipulating, our manipulations will result in disasters rather than in our survival.

Directing efforts away from this basic task of constructing a consistent picture of physical reality simply because of laziness, the inability to confront difficult questions, robs everyone who has put faith in the process to come up with answers that will allow the human race to move forward in the universe of an opportunity to contribute.

If we don't know the cause of motion, and we think we do, we will never attempt to uncover the cause of motion.

If we think we know the cause of motion when we don't, we'll die ignorant, never having had the chance to make the attempt to uncover the cause of motion and thus produce a perception of the universe that approximates realty, a perception of the universe that will allow us to extend our range of survivability in the universe.

3. MASS/GRAVITY IS BASED ON THE OCCULT NOTION OF STRAIGHT-LINE MOTION

THE MIND EVOLVED TO COMPARE REALITY AND RECALL

When we fail to comprehend things we see in reality, when we witness an event, a physical occurrence for which we have no explanation, we are uncomfortable.

The source of this discomfort is found in the way our minds work. The mind evolved to move our body safely through physical reality. To accomplish this, it stores pictures of physical reality in recall so that, as our body moves through physical reality, the mind can compare the pictures it receives from physical reality with the pictures it has stored in recall. If the pictures from both sources agree, then we can continue to move safely through physical reality. If the pictures don't agree, the mind stops operating, we feel discomfort and the discomfort stops us from moving through physical reality until we can form a picture of physical reality from recall that provides us with the recognition we need that will allow us to once again move with confidence through that physical reality.

This is how we keep from falling off cliffs.

The discomfort is also what causes curiosity, drives us to seek answers, makes us fill in the blanks of the pictures of reality that we form.

While the discomfort we feel when reality doesn't agree with recall is what keeps us out of harm's way, there are certain things about physical reality that aren't inherently harmful but raise questions none-the-less.

When someone throws a rock at us, we know the source of the motion of the rock, and we can put in our recall pictures that will allow us to deal with rock throwing and rock throwers. But when we see a rock drop, we don't have anything in our recall that would explain what is causing the rock to drop. We can put things in our recall that will allow us to avoid falling rocks, but we don't have anything to explain why the rocks fall in the first place.

Similarly, when we look at the sun moving across the sky, we don't have anything in our recall to explain why the sun is moving, and when we find out that it is the Earth rotating on its axis rather than the sun that is moving, we still don't have anything in our recall that would account for the Earth's rotation. When we realize that the Earth's rotation doesn't explain the moon's movement and become sophisticated enough to realize that the planets themselves are moving around the sun, we become totally stumped because there is no apparent cause for all the different motions, and therefore nothing to put in our recall.

And when we find out that light moves, we are completely lost.

The lack of an apparent force to explain the movement of the major physical occurrences in our existence produces discomfort when we become aware of them because we have nothing in our recall that would explain the motions and thus provide an agreement with those motions.

We can't go around in life feeling discomfort!

Because the motions are so very apparent, and so intrusive, they occur around us all of the time, we have to come up with something to put in our recall to agree with what we see or we would be discomforted to death.

WHEN THERE IS NO APPARENT CAUSE FOR REALITY, WE MANUFACTURE RECALL IN ORDER TO HAVE SOMETHING TO AGREE WITH REALITY

When we can't find answers to questions, we turn to the occult, the hidden world, the invisible domain obscured by the mists of the mind to obtain answers. When we are uncomfortable with the world we perceive, we are driven to create a world in our recall that can be used to agree with the world we see. When the world we see deals with the actions of people, the world of immorality, the evil men do, the unrequited agonies of endless wrongs, we create a world of religious belief, a structure that will give us answers that will allow us to avoid the discomfort of the unfair world we perceive.

However, when we are dealing with motion in physical reality we want objective physical answers for objective physical occurrences.

When the objective physical answers are not readily available and we are too lazy to pursue them, or when the pursuit becomes so overwhelming as to exhaust us, we settle for something else, not religious answers, but occult answers, agreed upon pomp and circumstance, repeated ritual, made-up words that will make us believe we have actually obtained objective answers.

Thus we create a force of gravity that is proportional to, and therefore a property of, matter to explain why objects drop. We don't know what gravity is, or how it is produced, how it moves through space, or how it affects objects to make them move, but we have Pavlov's answer for why objects fall, we say they fall because of gravity and we feel no discomfort.

We manufacture mindless answers to nonapparent reality that explain nothing but come to mind immediately when questions arise to quell the discomfort not knowing produces.

When we realize that the planet that provides us with life and safety is rotating at about a thousand miles an hour and moving through space at about sixty-seven thousand miles per hour, we make up words, angular momentum, or create visual confabulations, swirling masses of gas, occult crap both, to explain the motion.

And when we find out that light actually moves from one place to another, we puzzle away the discomfort by assuming light's existence away, saying that it is simply a wave, or maybe a wave particle or perhaps a particle wave, whatever, so long as it makes us believe that we have found an answer to the question of what force is causing the motion and therefore dispels our discomfort.

ALL MOTION REQUIRES A CURRENT FORCE AND CURRENT FORCE REQUIRES A PHYSICAL PROCESS

One thing that we should know as we become sentient is that when something moves, something has to be moving it.

We can, of course, attend a seance. We can sit around in a dim room with hands locked in a circle of trust and watch with amazement as the table rises in front of us, magically levitated by an occult force from the great beyond. We can snake our dollars hurriedly through our slathering drool to the medium creating the illusion, or we can fight our way out into the light and say, this is trash, things don't move without something moving them, I've been tricked, I don't know how, but I'm not going to turn my resources over to someone until they turn the lights on, come clean, explain to me in clear words, paint an understandable picture of what is going on.

But the light of reality is blinding and we are driven back into the darkness.

Dropping a gold coin on the risen table, the medium asks: "Where is the current physical effort moving the coin? You can't see the force that makes the gold coin drop, why demand a current physical effort to raise the table?"

The line between the light and the dark, comprehension and ignorance, is very thin. Lifting a rock requires an apparent force while there is no apparent force moving a falling rock. It requires effort to overcome the nonapparent force, and effort is visible.

The medium would have us believe that because there is no apparent force making objects fall, he can make objects rise with no apparent force. We know that it takes current physical effort to move something against the force holding the table on the Earth and we should know that it is the same force that makes objects fall.

We could therefore challenge the medium, fight against the forces of darkness peddling occult nonexplanations for actual physical reality. "Force is produced by a current physical process." we could insist. "Just because the force opposing our effort is not apparent doesn't mean that it is not the result of a physical process."

When we demand an explanation for the current physical process that is producing the current physical but nonapparent force that requires current physical and very apparent effort to overcome, we are standing on the side of the line that leads to light, that produces understanding.

If, instead, we deny the medium's claim that he can lift the table without a current physical effort but fail to challenge his statement that an object falls without a current physical effort, we abdicate our sense, we say the force holding the rock to the ground is not a force that takes effort, requires work, is the result of something physically happening and, crossing the line into darkness, begin to travel the path of ignorance that immerses us in the occult.

To demand physical effort to move something against the nonapparent force that makes objects fall and holds them motionless against a surface between the falling object and the source of the force, and to demand no physical process in the force's production, is stupid indeed.

If we fail to require a physical process in the production of the nonapparent force that makes objects fall, then we will be free to make up all sorts of occult crap about that force

SEPARATING THE FORCE THAT CAUSES OBJECTS TO MOVE FROM THE OBJECTS MOVING AS A RESULT OF THE FORCE

If we are to keep from slipping into occult explanations for force that confuses the force with matter, we have to isolate what we know about the force itself.

WHILE ALL MOTION REQUIRES FORCE, ALL MOTION DOESN'T REQUIRE THE SAME CURRENT FORCE

The first step in the descent into the occult belief that motion can occur without something physically happening, without a current physical process, is the observation that different objects require different forces to move. It takes more force to move a cubic foot of lead than it does to move a cubic foot of aluminum.

Motion becomes a product of force with the amount of force required to create the motion dependent on the nature of the matter being put in motion.

Clear thinking in the light of day would lead us to conclude that there was some force associated with the Earth that was holding the matter to the surface of the Earth. It takes effort to overcome that force and it takes more effort to overcome that force with respect to some matter than it takes with respect to other matter.

There can't be more than one force associated with the Earth so that the explanation for the difference in the force required to move different matter has to be found in the nature of the matter the force is acting on.

However, because it takes different force to move different matter, we are in danger of focusing our analysis on the matter rather than the force. Focusing on force when we are dealing with matter will make us confuse the two, and lead us to view the two as the same thing, view the force as the same as matter.

We might make the occult mistake of saying different matter exerts different force.

The fact that it takes different force to move different matter just means that more of the same force is keeping the matter from moving. It doesn't tell us anything about the force itself.

WHILE IT TAKES DIFFERENT FORCE TO MOVE DIFFERENT MATTER AGAINST THE NONAPPARENT FORCE, THE NONAPPARENT FORCE MOVES ALL MATTER THE SAME

We have to exert a current force to move matter because there is a force acting on that matter that is keeping it from moving. We can find out something about that force by seeing how it acts on matter that is not restrained from moving. We can drop objects and see how the objects drop in order to see how the force acts on matter.

Measuring the way matter moves when it is dropped and moving freely as a result of the force acting on it, we discover that the matter accelerates generally with the square of the distance over which it moves.

We also discover something else about the matter. We find that it doesn't make any difference whether we are dealing with lead or aluminum, or anything else. Everything moving freely in the force accelerates with the square of the distance over which it moves.

This again tells us something very important about the structure of the matter that the force is moving, how the force is acting on matter itself.

But we still aren't concerned with what makes up matter. We have to be very careful not to get sidetracked and confused. We don't want to run the risk of confusing the force with matter and thus thinking one is the other. We want to isolate the force that makes objects move toward the Earth and makes them stay there once they get there.

Knowing that the force causes an object to accelerate with the square of the distance over which the object moves, we can conclude that the force diminishes inversely with the square of its distance from the source, apparently the surface of the Earth.

WE KNOW TWO THINGS ABOUT THE FORCE

So we know two things about the force itself. We know that the force diminishes inversely with the square of its distance from the surface of the Earth and we know that the force is associated with the Earth because that's where released objects move.

We know nothing else about the force.

Nothing else!

We now have the choice whether we will accept an occult explanation for the source of the force or demand that the force be the result of something actually happening in physical reality.

AN IMPORTANT CHOICE

We can pretty much figure out that, if the force is associated with the Earth, then it is diminishing inversely with the square of its distance from the surface of the Earth because the surface area of the Earth is defined by the square of its radius. Anything emanating from the surface of a sphere would diminish inversely with the square of its distance because its distance is the radius of the sphere defined by its expansion away from that surface.

We know from our efforts overcoming the force that force is the result of something happening. We have to exert ourselves to overcome the force that holds objects to the surface of the Earth.

Can we expect anything less from the force that causes dropped objects to fall and stick to the surface of the Earth?

Do we conclude that there has to be something about the Earth, that the Earth has to be doing something, to produce the force that causes dropped objects to move toward its surface?

Or do we conclude that the Earth is doing nothing, that for this special case, where the cause of the force is not obvious, nonapparent, where it might take a little mental work to figure out the source of the force, do we conclude because we are too lazy or incompetent to figure out what's going on, that nothing is going on, nothing is happening to produce the force?

The answer to the question is important because if we decide to chose to view the nonapparent force as being the product of no effort being expended, then we will be helpless in our confusion, we will be incapable of separating the force from matter and we will step into the darkness, become comfortable with the ignorance that the force and the matter are the same, that one is a property of the other.

ARISTOTLE ORIGINATES THE BELIEF IN FORCE WITHOUT PHYSICAL CAUSE

Aristotle solved the problem of the nonapparent force making objects fall by assuming the force away altogether. He concluded that because the Earth was the center of the universe, anything under its influence simply gravitated to its surface because that's where the natural order of things required it to be. If a rock was lifted from the surface of the Earth, then it just sought out its normal position whenever the force that was keeping it from falling was removed.

Aristotle further claimed that thrown objects also moved of their own volition which made a fitting transition to his assertions about the objects in the heavens moving in circles around the Earth. These objects, Aristotle concluded, were embedded in the firmament, the famous aether, the medium through which Huygen's water wave light would be propagated several thousand years later, and as this aether moved around the Earth, it carried the heavenly objects around with it.

Thus, for Aristotle, objects on Earth moved with no apparent force, they were merely seeking their position at the center of the universe, and objects in the heavens moved as a result of a nonapparent force, the invisible aether. If something in the heavens came close to the Earth, then it escaped the nonapparent force moving the objects in the heavens and assumed its natural position at the center of the universe.

Aristotle set the precedent for two types of motion, one type for the movement of things on Earth and another type for the movement of things in the heavens above.

GALILEO KICKS OFF THE MODERN DECENT INTO THE OCCULT WITH ROTATIONAL MOTION

When Galileo was discovering the moons of Jupiter, scientists were called Philosophers and spouted Aristotelian precepts the same way empiricists spout eternal laws of motion, energy and matter, mindlessly. Copernicus earlier in the sixteenth century had delivered his deathbed work claiming that the Earth went around the sun. The Philosophers, pointing directly at reality, said, look, the sun clearly goes around the Earth, so shut up and die.

Galileo set about proving to these doubtless Thomases that the sun only appeared to go around the Earth, that it was in fact the Earth that was spinning and the sun that was standing still.

"If the Earth is spinning," came the chorus of philosophical certainty, "then what is making it spin? The fact that we aren't being thrown off the Earth into space is ample proof that the Earth isn't spinning."

Galileo, of course, couldn't answer the question, what made the Earth spin, because whatever is causing the Earth to spin is not apparent, just as what is causing objects to drop is not apparent. He could overcome the argument that we would be thrown off the surface of the Earth if it were spinning by claiming that whatever made objects fall and stick to the Earth's surface would keep us from being thrown off the Earth as well. But he couldn't come up with an explanation for what the force was that was causing the Earth to spin, if in fact the Earth was spinning.

However, if the Earth happened to already be spinning, and that spinning was occurring in nothing, then because there was nothing causing it to slow down, it would continue to spin. So with something having set the Earth spinning in the past, and with nothing slowing it down, it continues to spin right up to the present day. There was no question in the sixteenth century what set it spinning. God put it in motion as part of His creation, and He placed it in frictionless space so that it would continue right on spinning forever.

Therefore, the sun wasn't going around the Earth and Copernicus was proven.

"Not so fast." said the Philosophers. "We need some concrete proof the Earth is spinning on its axis."

Galileo went into his study and started rolling balls down inclined planes. A ball reached the end of the inclined plane and rolled across the table.

Galileo claimed that the ball rolled across the table after it left the inclined plane because the force the ball received from rolling down the inclined plane had been imparted to the ball and that if he created a frictionless path around the Earth, the balls would, as a result of that imparted force, roll around the Earth forever.

He called the imparted force inertia.

Galileo was taking the unknown force that makes objects fall and injecting it into the ball that was falling. He thus had the force coming out of the Earth into the ball where it would remain unless friction wore it off.

FRICTIONLESS MOTION ELIMINATES THE EFFECT OF THE FORCE THAT MAKES OBJECTS FALL

When Galileo claimed that the ball would roll around the Earth on a frictionless path forever once it had gained inertia from the force that makes objects fall, he was saying that the force that makes objects fall has no effect on the motion of the ball.

In other words, while gravity can cause objects to fall to the surface of the Earth, it can't cause them to slow down. Slowing down objects is the sole function of friction.

To make this point clear, Galileo pointed to a pendulum swinging back and forth. If, he stated, a pendulum could be constructed with frictionless parts, then once the pendulum was set in motion, it would swing back and forth forever.

This is really seriously deficient thinking. It's like saying that, without friction, a bouncing ball would bounce forever when in fact it isn't friction that slows a bouncing ball down, it is gravity. In like manner, gravity brings a pendulum to a halt regardless of how frictionless it is, and Galileo's ball would have about as much chance of rolling around the Earth forever in a frictionless path as Galileo would have of building such a frictionless path.

Here, where there is no chance of reproducing the effects claimed, the occult verges on outright fantasy and Empirical Science establishes its scientific facts.

Even though rocket scientists today claim that their satellites would orbit the Earth forever if space actually were frictionless, the Philosophers were not impressed with this twaddle.

Which is unfortunate because their continued skepticism led to the greatest misconception of all time.

GALILEO WHIPS UP A SYSTEM CLAIMING TO PROVE BELIEF TO BE FACT

In answer to continued doubts on the part of the Philosophers that the Earth was spinning because it contained inertia from God's hand, Galileo let the ball roll off the top of the table. It proceeded out into space until the force that makes objects fall caused the ball to start to fall just like a cannonball falls to the surface of the Earth.

A half century before, practical experimenter Tartaglia had gone out on the proving ground and shot off cannons until he determined that the optimum angle of the cannon for distance of the cannonball was forty-five degrees. Galileo, measuring the course the falling ball took in his study, computed the same angle, forty-five degrees.

Galileo then said that he had measured Tartaglia's reality in his study. Because he had been able to measure a reality not present in the confines of his study, Galileo claimed that, by mathematics alone, he could prove that which he could not in reality measure. Therefore, while he couldn't in reality measure that the Earth was turning on its axis as a result of inertia, the fact that the ball obtained inertia from the inclined plane proved that the inertia existed and that the Earth was turning on its axis because of this proven inertia.

The Philosophers duly impressed, they were, after all, as confused about force, matter and motion as Galileo was, asked Galileo for objective verification that the Earth was spinning. Galileo, knowing the tides worked just as if the Earth's rotation were causing them, pointed to the tides as confirmation that the Earth was spinning in space.

Unfortunately, the friction from the tides would have long since overcome the Earth's inertia, so that was the end of that, unless one wishes to discuss *vis viva*, the force Galileo created to imbue matter with life thereby allowing it to move without apparent force, a religious inertia that moved forward in Empirical thought of its own accord through the nineteenth century.

Copernicus remained unverified even during the seventeenth century when it became something of a test of the scientific competency between nations to come up with proof. Robert Hooke, the underpaid and thus obsequious employee of the British Royal Society, under undue pressure from his employer, even attempted to measure the predicted displacement of stars measured on opposite sides of the Earth's orbit as proof that the Earth had opposite sides to its orbit and thus moved around the sun. His claimed parallax measurement, eventually drastically disproved as delusion (to be charitable), didn't stop future empiricists from using parallax measurements to measure the nonmeasurable distance to the stars, just as Galileo's failure to demonstrate inertia didn't stop empiricists from not only firmly believing in its existence, but in believing also Galileo's absurd claim that beliefs could be proven by mathematics alone to be fact, that if something is computable in the mind, it exists in reality.

NEWTON TRIES USING GALILEO'S SYSTEM OF PROVING BELIEF TO BE FACT TO PROVE HIS BELIEF THAT THE FORCE THAT MAKES OBJECTS DROP IS THE SAME FORCE THAT CAUSES PLANETS TO ORBIT

In the fourteenth century, Buridan came up with the maxim that impetus would endure if unopposed. Between the time of Buridan and Newton there was some pretty strange stuff espoused about motion, especially by the ever influential Huygens, of water wave fame, who noted that circular motion would require centripetal acceleration. Huygens had turned light into a wave to assume it away and thus avoid explaining the source of the current force that was causing it to move. "Centripetal acceleration" was just another way of assuming away current force with the use of words, constant circular motion requiring only the constant application of a current force.

It took Newton to crystallize everything into a few neat laws, the first of which was the occult notion that an object will move in a straight line unless a force acts to change that movement, occult because it assumes away a cause for the straight-line motion.

Why did Newton need a law that something would move in a straight line unless a force acted on it to change that movement?

Newton was attempting to prove that the force that caused objects to fall was the same force that was involved in planetary orbiting. If he could show that the planets moved in response to the same force that caused objects to fall, he could claim proof for the occult notion that the force was a property of the matter toward which the objects fell.

Newton's problem was to connect the force to the matter, and the only way that he could do this was to make the force proportional to the matter of which it was a property and then show that the moon moved with respect to the Earth as if the force was produced by the amount of matter in the moon and the Earth as diminished inversely with the square of the distance between the two, the mathematical rate at which falling objects accelerate, the acceleration conveniently satisfying Huygen's need for acceleration in circular orbits (which is to say, oh my, even though this is taught in every classroom on the planet, I'm so embarrassed to put it in print that I have to force myself, *the moon stays in its orbit because it gains speed by falling toward the Earth!*).

Newton could compute the amount of the force in the Earth and the moon by computing the amount of matter each contained by volume. However, this only gave him half an equation. For the other half, he needed to know the opposing force, the amount of force the force of attraction between the Earth and the moon was overcoming.

This force was found in the moon's momentum as it circled the Earth. If he assumed that the moon wouldn't be circling the Earth but for the force of attraction between the two, then it was very simple to compute the amount of force found in the moon's momentum. It was a product of its size, the amount of matter it contained, and its speed, how fast it was going.

To obtain the second half of the equation, Newton needed to assume that the moon would be moving in a straight line but for the fact that it wasn't, the thing he was trying to prove, that the force that made objects drop to the surface of the Earth, which diminished inversely with the square of its distance, was proportional to the matter in the Earth and the moon because the force that amount of matter produced, reduced inversely by the square of the distance between the Earth and the moon equaled the amount of force required to bend the straight-line motion of the moon into its circular motion.

By showing that the force was proportional to the amount of matter producing it, Newton could declare that it was a property of that matter.

The whole scheme failed to balance with respect to the moon, and when applied to the planets, failed abysmally.

Faced with this failure, Empirical Science was forced to accept the inverse square proposition as law, as opposed to a hypothesis that had to be proven, or even a proven hypothesis, and now uses it to create matter wherever it can measure motion, and some places, such as the hearts of nebulae, where it can't.

But where did that leave the occult concept of straight-line motion?

NEWTON CALLS GOD INTO OF THE MISTS

Newton was well aware of the problems involved in his use of mathematics to prove that which can't be measured in reality, that the moon and the planets move in response to the same inverse square law that makes objects drop. All the measurements of the moon's motion made in Newton's lifetime failed to prove the belief, with Newton carrying the hope to his grave that when measurements got accurate enough he would be vindicated.

However, before his lifetime, during his lifetime, and for all periods after his lifetime, there was no empirical basis for the statement planets move in a straight line but for the force of gravity bending that straight-line motion. Whewell, using Newton's mathematical hocus pocus in an attempt to craft some sort of rational statement for Galileo's method of turning beliefs into fact kept running into the problem of justifying the statement. While never coherently doing so, creating in the process a sort of inductive regression where the empirical truth was up there somewhere in a hierarchy of empirical statements, Whewell knew that Newton had found the secret of the universe, so he just kept on rephrasing the law, knowing its truth lay in the sense of all mankind that it is thus so it must be so.

But to explain where the force came from that put the planets in motion was a different matter. Newton knew the source of that force. It was God.

And, responding to the instinctive feeling that Galileo was wrong, that the gravity exerted by the matter overcoming the straight-line motion of the heavenly bodies would eventually erode that motion, or perhaps foreseeing the truth of Huygen's belief that light was like a water wave after all, and the aether through which the waves were propagated would slow the planets down, Newton claimed that God came back periodically to minister to what He had created, thereby keeping a planet's speed up to snuff to overcome the ever-present gravity He had created at the beginning of the universe.

GOD EMBARRASSES NEWTON'S HEIRS

We can only imagine the chagrin, then, when it was discovered that there was no God, and if there were, He certainly wasn't one to come back to this little corner of the universe to give the planets a periodic boost, and even if She was one to come back, Her very existence was an embarrassment to a science that prided itself on producing objective explanations for physical reality rooted right in the physical reality itself.

Thus, motion as a result of God's hand was out of the question.

What a pickle to be in, having a bunch of laws that were unquestionably true, objects moved in straight lines, gravity was proportional to and therefore a property of matter, space was frictionless, and no explanation for the foundation of those laws, that things moved without force. Without something setting the moon and the planets in motion at some time in the past, there was no straight-line motion for gravity to overcome.

Empirical Science, however, never at a loss for making up ad hoc absurdities, came up with one of its best.

If, Laplace reasoned, the solar system condensed out of a swirling mass of gas, then the motion of the planets would be imparted to the matter making up the solar system by the swirling mass of gas.

And, to boot, with a little imagination, planetary rotation, which Newton had ignored because it wasn't computable in his mathematical scheme of things, could be ascribed to the same source, the swirling of the mass.

Of course, if there was a mass of gas, and the gas was made up of matter, and the matter had mass, and mass had gravity, then the mass of gas wouldn't swirl, it would just go clump in the night of space!

But, when Empirical Science seeks to explain reality, it can't let a little thing like reality intrude. Reality, while based on, and obeying Empirical Science's mathematical processes, is not reality if it conflicts with the mathematical outcomes of those processes.

And there is nothing more mathematical than a swirling mass of gas. Once the random motion in a jar full of gas molecules can be explained, it's all downhill, the direction most empirical conclusions lead.

Even the Smithsonian's wall-size chart showing the origin of the solar system has a big question mark about this swirling mass of gas. But when there is no other explanation, any explanation is acceptable.

SO WHAT DOES THAT LEAVE US?

The swirling mass of gas is required only because straight-line motion is required, and straight-line motion is required because otherwise there would be no basis for proving that gravity was proportional to, and therefore a property of matter, the explanation for the original nonapparent force, the force that causes objects to fall.

But gravity was never proven to be proportional to matter, Newton's use of volume being incapable of predicting motion.

And even assuming that the force could be somehow connected to the matter, the conclusion that it is a property of matter doesn't necessarily follow. It could, for instance, be the result of something the matter is doing.

The whole enterprise was poisoned from the beginning, for reasons more cogent than the claim that volume was not necessarily an indication of the amount of matter present, but once the enterprise was undertaken, its consecutive failures had to be cemented over until the empiricists doing the cementing had no concept left of the original structure.

And yet, the whole stagnant enterprise is based on a belief in straight-line motion, in the necessity that, because of pure laziness and incompetence in the attempt to figure out the cause of motion in the first place, motion has to be assumed away, its cause placed somewhere out of reach in the past.

Once we base an enterprise on belief, it doesn't make much difference whether we source that belief in God or swirling masses of gas or gravity wells, we're just sitting around making up stuff and then agreeing on which stuff deserves belief.

We don't know what causes objects to fall, we don't know what causes planets to rotate on their axes, we don't know what pushes planets around in their orbits and we don't know what causes light to move from one place to another. We don't even know what holds the atoms together to form the paper on which these words appear and we know nothing about the nature of the light that allows the words to be read or about the electricity that produces the light that allows them to be read.

Our ignorance is stunning, simply stunning.

Making up occult explanations for the hidden causes of motion is understandable. It's a simple defense to the discomfort that not knowing produces.

But automobiles and airplanes, rockets and computers were not built by people omming around a table waiting for it to rise magically into the air.

Technology is created by engineers who ask hard questions about physical reality, then sit down and, through a process of trial and error, come up with ways to alter that physical reality.

When we believe we know something, when we think we know all the answers, then we won't attempt to find out what the actual explanations are.

An understanding of motion is at the basis of all technological enterprise.

Believing in nonexplanations for motion or assuming motion away, believing that it occurred in the past, or that it is inherent or that the thing moving doesn't actually exist limits how we will tackle technological innovation.

While our ignorance is stunning, the fact that we docilely accept absurdities as explanations for the forces that rule our lives is downright stultifying.

Empirical Science has made us the Cargo Cult of the universe!

4. THE SOLAR SYSTEM DOES NOT OPERATE ON THE BASIS OF HISTORICAL FORCES

NAVIGATIONAL NECESSITIES

Today we can precisely fix our position on Earth by bouncing signals off satellites. We can rent a car which contains a computer screen depicting a map of the city we are visiting. The car's Global Positioning System will locate the car on the map so that we need never be lost.

A hundred years ago, when a ship left port, its position was up for grabs, estimated only by the dead reckoning of its Captain and his ability to determine position from observing the starry lights in the night sky.

This was how we navigated throughout all but the most recent history. The inability to navigate kept navigation at a minimum.

Prior to the fifteenth century, the European demand for the spices from the orient was satisfied by Indian merchants riding the Monsoons to the Moluccas, the so-called spice islands east of Java, and returning with the spices through the Straits of Mallaca to the Malabar Coast where Arab traders transported them in their dhows on the ancient coastal trip to Egypt. The Merchants of Venice then brokered them within the confines of the Mediterranean or overland to European customers. These puddle jumping trips required little navigation because the ships hardly ever lost sight of land.

The spice trade, being a monopoly, was incredibly profitable and anything that is profitable attracts industry and invention. The industry and invention in the case of the spice trade was found in the Portuguese Prince, Henry the Navigator, who believed that if a way could be found to sail down and around the African Continent and out into the Indian Ocean that the Venetian choke hold on European throats derived from the preservative nature of the spices could be broken.

The Muslims, who had taken Constantinople and therefore played a significant role in the spice trade, also controlled the northern part of the African Continent, and Portuguese sailors had to head for deep ocean in order to conduct the exploratory voyages necessary to make their attempts a reality. In deep ocean, a ship's position had to be fixed by the stars. Prince Henry was responsible for much of the navigational equipment that was to become a necessity in the following centuries as trading nations fought for trade routes in the endless oceans of the world.

For thousands of years prior to Prince Henry's efforts to reach the Orient, the Ptolemaic universe ruled the heavens. The sun went around the Earth, as was attested each morning by it coming up in the east, traveling across the sky from east to west, and then setting in the west. The stars also went around the Earth, as did the planets.

Beliefs hold where there is no reason to question them, and the daily lives and economic relationships of the world prior to the fifteenth century provided no reason to question the belief that the Earth was the center of the universe. Many pointed out that it was the Earth that traveled around the sun, Aristarchus being the most vocal, but where it makes no difference, what people believe will dictate the reality that people see.

When serious navigation began in the fifteenth century, belief started to disagree with reality. The investors who backed the captain whose belief that the Earth changed its position among the stars because it traveled around the sun and who therefore brought back his ship laden with spices were much more receptive to believing the Earth actually went around the sun than the investors who backed the captain who believed that the stars went around the Earth and therefore ended up wreaked on a desolate section of the African Coast were to continue in their belief that the sun traveled around the Earth.

Economic success drives the mind to accept facts in reality that disagree with the facts people hold firmly in their recall as beliefs. Thus the stage was set for Copernicus a hundred years after Henry's thrusts deep into the Indian Ocean to formerly posit that the solar system was heliocentric, with the planets moving around what was believed to be a fixed sun in the center.

In the ensuing centuries, the necessity for navigational certainty led to the positing of theories that could be used to predict the position of the planets and the stars in the skies and therefore increase the accuracy of navigation. Knowing where a star or planet should be in the sky was half the battle in finding out at what latitude the ship was located, the other half plotting the ship's location from the position of the star.

Much labor was put into the production of ephemerides, tables positioning heavenly bodies over specific periods. These tables by necessity involved viewing the motions of the bodies as that motion had happened and then determining by inference where the stars and planets would be in the future. This is not the act of predicting, but rather the process of recording experience that is expected to recur in the future.

Sailors would then take these tables on their voyages and note on them any actual discrepancies with reality. Because drawing inferences also involves basing perceptions of continued movement on theory, theories underling the ephemerides were constantly being reevaluated.

A HIDDEN ASSUMPTION

The theory that could best produce accurate ephemerides was considered to be a theory close to reality.

These theories were not really theories, but rather conceptions about what the facts might be. The Ptolemaic system survived through Kepler's predecessor Brahe. From the pure Ptolemaic process of the planets and the sun all moving around the Earth, observational reality had forced changes until Brahe's solar system was reached, a solar system in which all the planets moved around the sun and the sun moved around the Earth.

The intervening degrees of conceptual evolution created a system of eccentric circles and epicycles together with equants, imaginary points around which the eccentric circles, epicycles and equants rotated. The heavens before Copernicus were made up of a changeable congeries of flying pinwheels and double rotating ferris wheels as concentric circles and epicenters were altered to what the builder of the particular ephemeris believed to be the situation.

We have to be very careful about how we view what was going on here.

What goes around what in the solar system are simple facts. We just didn't know what the simple facts were because we couldn't get far enough away to take in the full picture. Galileo's discovery of the moons of Jupiter revealed what the solar system probably looked like at a distance, and looking at the Jovian system through a telescope could convince all but the most diehard fanatics of the Copernican reality, but any description of how the facts were ordered was simply an attempt to draw a picture of reality.

The builders of the ephemerides were constructing templates against which they could measure observed reality. They were taking what they saw, converting it into a picture of how they thought things were, and then saying if things are like this, then the positions of the planets will be thus and so next week, next month, and even next year.

Underlying all of these efforts, however, were several assumptions hidden from conscious consideration. Any process that used the past to determine the future, while not prediction, was based on the prediction that what has gone on in the past will continue to go on in the future.

And when we think that something that has gone on in the past will continue to go on in the future, then we believe that what we are seeing is an orderly, unchanging process.

We are used to the seasons following one another, we are accustomed to the moon traveling around the Earth and the sun always rising. These things are orderly, repetitive, without change.

Therefore, it isn't surprising that any template created to describe the movement of matter in the solar system would forever fix the motion of the planets in space, hold them to orderly courses which were both repetitive and predictable.

More important to the process of making irretrievable mistakes, once we have the picture, we say, ah ha, I have the picture. We say, I've got it, I finally know how the movement of the planets in the solar system happens, the planets go round and round the sun on a regular basis.

And because the picture is correct, the planets do go round and round the sun on a regular, orderly basis, we let the hidden assumptions enter our thought processes. The planets go round and round the sun on a regular basis in accordance with the template so that we can predict where the planets will be for any period into the future and we can compute where the planets have been at all times in the past.

We had to assume orderly motion to construct the template that would allow us to construct an ephemeris that could show future positions and once we got the template that uncovered the huge mistake, that the sun didn't go around the Earth, we accepted all aspects of the template and imposed its order on the movement of planets.

A PICTURE OF HOW THE SOLAR SYSTEM OPERATES DOESN'T TELL US ANYTHING ABOUT HOW THE SOLAR SYSTEM OPERATES

The reason that we have to be very careful to separate the picture we are constructing of the factual make-up of the solar system and theories about how the solar system operates is if we ever hope to make sense out of what we see, the facts in our reality, we are going to have to be very careful to avoid believing theory to be fact. If we go around believing theory to be fact, then we are no better than our forebears who had a backwards picture of the facts, who believed the sun went around the Earth.

There is no difference between believing wrong facts and believing theory to be fact. Neither exists in reality.

In our self-congratulatory frenzy at finally getting a clear picture of the operation of the solar system, at seeing the factual reality we had so long sought, we naturally believe that we know how the solar system operates.

After all, we can see it operating in our minds as the planets move around the sun and the moons move around the planets.

Isn't this, after all, the operation of the solar system?

Isn't knowing that the planets move around the sun knowledge of an operation?

Doesn't operation mean what it means, an act, process or way of operating? The solar system doesn't operate by the sun and the planets going around the Earth, it operates by the Earth and the other planets going around the sun.

We now know how the solar system operates.

When something is in motion, we have a tendency to stop at the surface. When a child learns how a clock works, he says the hands move and when the big hand points to the one and the little hand points to the five, it's five o'clock. He doesn't have any knowledge about the springs or crystals, movements or electronic circuits that keep time pieces operating.

There are two levels of operation, one what we see on the surface and the other what is actually happening.

Knowing that the Earth and the other planets move around the sun doesn't tell us anything about how the solar system operates. The word operate is derived from work, and work implies force, something happening to make something do something. An operation isn't an operation without a doctor cutting and slicing, performing current work that consumes energy.

Knowing that the planets move around the sun doesn't tell us anything about what is causing that movement.

What don't we know about how the solar system operates?

We don't know what causes the planets to move around the sun, and the moons to move around the planets, and we don't know what causes a planet or moon, or the sun for that matter, to rotate on its axis.

Obtaining the picture of how the matter in the solar system moved told us nothing about how the solar system operates!

And yet because we believed it did, we sat back happy, self-contented, brain dead.

RELATING THE FORCES THAT MOVE THE PLANETS TO THE HIDDEN ASSUMPTION THAT THE SOLAR SYSTEM OPERATES ON A REGULAR BASIS

Overlooking explanations for the forces that are at work to cause the movement we see in the solar system is such a massive display of ignorance that we have to wonder how such ignorance came about. After all, even though we know the sun is not moving across the sky each day, the fact is, it appears to move across the sky because the Earth is rotating on its axis. Not knowing, not even asking why the Earth rotates on its axis means that we are so blind we don't even know why the sun comes up each day.

The answer we satisfy ourselves with, that angular momentum is conserved in closed systems doesn't tell us anything at all, is worse even than going around believing that the sun travels around the Earth because we are knowingly pulling the wool over our own eyes.

And when we actually see something going around the Earth, as we watch the progression of the moon across the night sky from hour to hour, to not know what is making it move, to not even ask why it is moving is colossal stupidity of the grossest proportions. The answer we satisfy ourselves with, that it is moving because of its momentum doesn't tell us anything at all, tells us less, if that's possible, than blathering about angular momentum when we see the sun rise.

The reason that we can walk around in such mammoth ignorance of the everyday realities we live with is found in the template we accepted, frozen in time when we finally got the picture of the movement of matter in the solar system right, not in what the template demonstrated, but rather what it implied.

With matter conforming to the movement of the template, then there is really no reason to explain what the forces at work moving the matter are because whatever the forces are, they are contained in the matter doing the movement and have been contained in that matter since the matter was put in motion.

For instance, Galileo attempted to address the rotation of a planet on its axis by saying that an object, once set in motion, would continue in motion unless friction caused it to stop moving. He called this property of matter inertia and claimed that because the Earth existed in frictionless space, its inertia explained its rotation. It wasn't necessary to show where the force came from because God created the universe and there was no basis for concluding that He created it without the motion we see in the matter He created when He created the universe.

Of course, space is anything but frictionless, and using a historical force for current motion doesn't explain anything, but if we are ever in doubt about the apparent explanation, all we have to do is point at the sun coming up every morning, the same sun our forebears used to claim that the sun moved around the Earth, and say, the Earth is rotating because of a historical force, because angular momentum is conserved in closed systems.

When Newton directed his attention to the movement of the moon around the Earth he concluded, using much the same tried and true reasoning that Galileo used, that the moon was moving as a result of the momentum it received when God created the universe, the tried and true referring to the reasoning rather than the cause of movement. If a misconception can exist uncontradicted for a century, then it is a pretty safe misconception to hang your hat on.

The only problems with Galileo and Newton's explanations for motion in the solar system, that it was a result of historical forces, was the explanation for that historical force, God. By the nineteenth century, God as a source of scientific force was definitely out of play.

As a result, the ubiquitous swirling mass of gas was created to explain both planetary rotation and orbiting, with the planets condensing out of the gas cloud with the swirling motion of the gas cloud imparted to the forming planets. This had the advantage of keeping the force historical while rendering the source invisible.

In this flight of fancy we were putting a template on top of a template. Remember that the template we created to produce a picture of the movement of the planets was perfected by using the actual movement of the planets. We took facts and created a template to allow us to visualize the facts. In the case of the movement of matter in the solar system, the successful template approximated the facts, and coincided with one central fact, what moved around what.

We were using facts to uncover a fact.

When we created the template of the swirling mass of gas, we were using facts, planetary rotation and orbiting, but the template we were creating was about facts that might or might not be true. We can't see the force that causes the motion in the solar system like we can see the motion of the matter in the solar system, so we don't know what it is. In fact, because we can't see the force, we assume it isn't there.

Because we can't see the force, we are free to make up any template for it we chose, and there will never be anything to contradict that template. We are dealing in pure theory, and theory is never fact.

By producing a template made up of a swirling mass of gas, we think we have created the same reality we produced when we created a template for the movement of the matter in the solar system, so we enshrined the assumptions of the original template, uniform, predictable motion, in the new template.

When we ask what causes the planets and moons to orbit and rotate on their axis, we say they orbit and rotate because of the force that was imparted to them by the original swirling mass of gas out of which they condensed.

And in doing so, we ignore the fact that matter only moves as a result of current force, therefore encasing our ignorance in stone.

THE CONSEQUENCES OF ACCEPTING HISTORICAL FORCE AS THE SOURCE OF THE MOTION OF MATTER IN THE SOLAR SYSTEM

Only the truly witless, those ignorant beyond redemption, those so sure that they know what makes planets and moons rotate and orbit simply because they know that planets and moons rotate and orbit, can feel comfortable going to any lengths to discredit pictures of the solar system that disagree with their made-up pictures and still believe that they can maintain their integrity.

When you know something can't be wrong, then you know you have to be right in discrediting opposing pictures and it therefore doesn't make any difference how ridiculous your statements are in discrediting opposing pictures, because any picture opposing your picture has to be wrong.

The swirling mass of gas template used to hide the fact that we are ignorant of the current force that causes movement in the solar system dictates that the orbits of the planets and the moons cannot vary.

The motion that Newton assumed in orbiting planets and moons, straight-line motion maintained by momentum, is bent into circular motion by the only current force in the universe, the force of gravity. Thus, the matter in the solar system, other than the sun, appears in orbits created by the straight-line momentum being exactly counterbalanced by the force of gravity which is bending that straight-line motion into the circular motion that describes the orbits.

Because the force that put this matter into motion is a historical force, it is no longer present, then there is nothing to change that motion. With motion caused by a force that is no longer present, and the assumption embedded in the thought processes that the motion we see has always been and always will be, there is nothing other than the gravity that is proportional to matter to change the motion of matter in the solar system and because matter can neither be created nor destroyed, the amount of matter never changes and the motion can never change.

Thus, when Immanuel Velikovsky, a nonliterate in the workings of the solar system, uncovered archeological evidence that pointed toward the possibility that Venus, in recorded history, crossed the orbits of both Mars and the Earth, and witlessly published it as scientific, men of certainty such as Carl Sagan felt no restraints in discrediting him in the grossest possible public manner.

On the one side, Velikovsky had gone out and dug up ancient astronomical records that supported what he asserted.

On the other side, Sagan represented the consensus which asserted that the solar system was set in motion by historical forces, that there was no known force that would allow the planet Venus to move about in its orbit, after all, historical forces cannot cause changes in current motion, and that any records of Venus moving around in any orbit other than her current orbit, had to be either forged or misconstrued.

In short, one side was dealing with fact and the other with theory, and where theory disagrees with fact, theory always prevails. Sagan, firmly believing in the template, the immutability of the solar system, demanded empirical proof of Velikovsky's claim and poor Velikovsky, put on the defensive, kept coming up with proof not realizing that, because facts can never alter beliefs, believers will continue to spout any stupidity that comes to mind to argue away facts opposing those beliefs.

Because of assumptions made when we reveled in obtaining a correct picture of the movement of matter in the solar system, we have crippled our picture of the solar system, blinding ourselves to realities that exist right in front of our eyes. Like the believers in the Ptolemaic system, facts will never change our beliefs. Unlike the Ptolemaic system, where measurable fact existed to change belief, there is no central fact to change the belief in the template of an orderly, unchanging solar system because there is no fact to support such a belief to begin with.

NO ONE HAS EVER MEASURED REGULAR MOVEMENT IN THE SOLAR SYSTEM

Newton, in deriving his mass/gravity scenario which produces the belief that we understand the forces at work in the solar system as well as in the universe as a whole, recognized that there were constant discrepancies in the movement of matter in the solar system. It doesn't take a rocket scientist, whose predictions of performance are made difficult by the assumption of uniform motion, to see the irregular motion of the movement of the moon. Entire books have been written trying to reconcile the irregularity of the moon's motion with the swirling mass of gas template of planetary operation. The moon is even, though delusionally so, measured to speed up and slow down, something that couldn't be a product of historical forces.

There are so many discrepancies between theoretical and factual reality with respect to the motion of the matter in the solar system that the unexplained motions are categorized as perturbations for easy filing and forgetting.

Newton felt that to maintain the solar system's equanimity, to keep the planets moving in their prescribed paths, God had to come around and do a little touch-up work periodically. Halley computed that the discrepancies he measured in Jupiter's orbit were causing it to contract by a degree every thousand years while Saturn's orbit was expanding by twice that amount over the same period. Newton died in despair of Halley coming up with measurements of the moon's orbit that would show that it was operating in accordance with an inverse square law, the moon never operating in a uniform enough manner to confirm what its actual orbit was.

Laplace, who dreamed up the swirling mass of gas as a replacement for God, proved mathematically that the gravitational pull between Jupiter and Saturn caused their orbits to contract and expand in nine hundred year cycles. Not being capable of disproof, this fantasy became further proof that Newton's static solar system was the correct picture, leading to Laplace's famous declaration: "Such has been the fate of Newton's brilliant discovery, that each difficulty that has arisen has become for it a new subject of triumph, a circumstance that is the surest characteristic of the true system of nature."

Of course, the first difficulty was that Newton's theory that the force that makes objects fall to the surface of the Earth was the same force that bent the straight-line motion of planets into circular orbits was never proven leading perforce to its adoption as law and then its use as law to claim knowledge as to what was in the center of the planets, safe claims because their accuracy could never be checked.

If planets and moons don't move on a regular basis, then how come they have stayed in orbit for the five billion years we delusionally think they have been around?

In showing mathematically that Jupiter and Saturn have repetitive orbits that makes them move back and forth from each other in nine hundred year cycles, Laplace had to assume that the variance in a planet's motion had to have boundaries. That's how you mathematically prove fantasies about reality, you assume all sorts of constants so that your mathematical equations balance, and if they don't balance, it isn't because they don't describe reality, it is because the wrong constant has been chosen, the right constant has been chosen but it hasn't been assigned the right value, or there is more than one constant. That's why computers are so valuable in determining what reality is, they can handle more calculations with less constants during any particular funding year.

Perhaps we could just say that the motions of the planets are all over the place, and as a result cancel each other out.

That, however, has no scientific validity because it is not a quantized statement that can be used to predict outcomes.

Another solution to the instability of the solar system would be to say our hidden assumptions were simply wrong, that the solar system didn't operate on historic forces, it operated on current forces that varied on a daily basis. Because the production of the force was steady on an overall basis, the force was actually holding the system together against variances that tended to cause it to come apart. We could use the fact that the force did a good job of producing order out of chaos as a clue to the nature of the force itself.

However, to take this approach would require the use of the mind. While we may well have minds when we are trying to figure out what to order for lunch, we don't in reality have minds because the mind has never been cut out of the brain and put in a bottle. Whatever it is that does our thinking for us is not something that we should give much credence to in trying to figure out what is going on in the universe. Forget the fact that it is what we have to make sense out of reality, whatever it is, it's too subjective, too flawed.

Better to make mindless conclusions such as historic force and uniform motion which keep us from thinking than make assumptions that something is actually happening which would require us to do a little thinking to figure out. Thinking is work, and work is not particularly useful when we can produce mathematical systems to think for us and to conform reality to, and especially when we can produce computers that can do mathematical computations even better than we can.

Perhaps we could just handle the solar system's instability the way we handle the tides. Everyone knows that the tides are caused by the rotation of the Earth and the movement of the moon and the rotation of the Earth with respect to the sun.

No tide has ever been predicted on the basis of computing the movement of the moon, the Earth or the sun, but that doesn't seem to bother anyone because everyone knows that the tides can be predicted by those motions. We can just say, well, we know the tides don't conform to the movement of the matter causing them, but because we know they do, we just haven't figured out how to go about predicting them. We'll get busy on that next week. In the meantime, if we want to know when tides occur, we can just go to where they are occurring and measure them for a day or two, and we'll have the rhythm of the thing.

But the planets don't wash up on any shores, so this logic seems to fail us in explaining the instability of the solar system.

Woe, oh woe is us, we lament. How the Swirling Mass of Gas has rained down troubles down on our bowing heads.

PUTTING IRREGULAR MOVEMENT ON A REGULAR BASIS

We can't analogize the movement of matter in the solar system to the tides, although we ascribe the tides to the movement of matter in the solar system and we can't use our minds to counter the fact that there is no regular movement in the solar system.

Yet we have to say something because the irregular motion is constantly measurable and thus constantly coming up.

What we need is a constant that would allow us to be able to compute the regularity of the irregular motion so that we could make some statements about the boundaries of the motion and thus find out the laws that govern it.

The way to do this would be to produce a mathematical portrait of the orbits of the planets at a hypothetical point when they condensed out of the swirling mass of gas. Then the orbits of the planets could be computed forward in time to the present from this initial condition.

But this process won't provide anything to compare the orbits with, so the way to accomplish a comparison would be to produce a second set of initial conditions slightly different from the first set of initial conditions.

This, of course, is precisely the situation we have in the solar system. With only mass to bend straight line motion, as the two sets of initial conditions were tracked over billions of years, the precise variance set as initial conditions should continue throughout the tracking.

What was found, duplicating reality this way on a computer, was that an initial difference as small as one ten billionth of an orbit, which is about how far the Earth travels in one three hundredth of a second, resulted in a variance so great that it would be impossible to specify where the Earth would be with respect to its sister initial condition after a hundred million years.

Thus, it was clear that whatever the unknown boundaries at work were, boundaries such as Laplace had used to mathematically prove that the measured change in the orbits of Jupiter and Saturn operated on a nine hundred year cycle, they were still at work.

The only job was to figure out what the boundaries were that allowed the chaos demonstrated in the solar system by duplicating the initial conditions of its operations and checking them against an almost identical set of initial conditions to maintain the uniformity of movement that existed.

What we have done is to assume that there is uniform motion in the solar system as a result of a historical force, and when we found that the uniform motion we had assumed wasn't uniform at all, we created a computerized process that would show us how irregular the motion was so that we could create assumptions that would demonstrate why the irregular motion we measured was actually regular motion.

It's lucky that we aren't depending on this silliness to navigate on the far oceans of the world because we would fall off the face of the Earth.

WHO IS WE?

When we first sent satellites up to orbit the Earth, we expected them to stay in their perpetual Newtonian orbits forever. When they didn't, when we found that they stayed in orbit, or fell from orbit with no rhyme or reason, we looked for the cause of our satellites' failure to meet our laws of nature. Noting that solar flares seemed to produce some drastic shifts in the stability of satellite orbits, we looked back to Galileo for the answer we sought. We knew that the satellites would continue to move in a straight line but for the force of gravity, and we knew that the force of gravity wasn't increasing because we had made up a law that stated that gravity was a property of and proportional to mass and the amount of mass in the Earth wasn't increasing with the activity on the surface of the sun.

So, if mass wasn't increasing, the only other reason for the satellites to change orbit was if there was friction being produced in space which hadn't been there without the solar activity.

What could cause an increase of friction in space?

The only thing around was the Earth's atmosphere, so the solar flare-ups on the sun must be affecting the weather in such a way as to cause the atmosphere to expand into space causing friction which slowed the movement of the satellites.

By golly, there it was, the empirical evidence needed to understand why our laws didn't apply to a satellite's orbit. Can't get much more scientific than that, unless its the conclusion that the tides are caused by the moon. If the tides weren't created by the moon, then the friction their movement would cause would have slowed the Earth's rotation down. The Earth is still turning so the tides have to be caused by the moon. The falling satellites were absolute confirmation of an expanding atmosphere because if the atmosphere wasn't expanding, the satellites wouldn't be falling.

When we sent the first probes to the moon using our laws of motion, they came nowhere near the moon. It was only when we started recomputing the courses on the basis of actual reality, on how far off the probes were from the made-up reality computed by our laws of motion, that we started to hit the moon. And, because we had no idea where the moon was going to be when the probe got there, we had to evidence eternal vigilance on a visual basis to make sure the probe was not going to arrive when the moon didn't.

We spend a lot of time marveling at how we can send a space probe to the outer planets with such accuracy, but the science of orbital determination, where teams gather around daily to examine pictures taken from the probes, delays in radio signals sent from the probes, and beams of one sort or another bounced off the probes in order to figure out where the hell the probes actually are, belies this accuracy.

The entire trip of anything in space is a continuous process of course correction based on trail and error and has absolutely nothing to do with our hoked-up laws of motion!

The simple fact is, reality, the movement of matter in the solar system, or for that matter, the movement of falling objects on Earth, has nothing to do with our made-up laws about what is causing that motion.

Bacon referred to the cause of motion as the "secret motion of things" and he said that this is the goal of our search for understanding, to come to an approximation of the explanation for that motion, falling objects, rotating and orbiting planets and moons. What causes the motion cannot be seen, and the only thing with respect to the causes that can be measured are its effect on the matter it is moving.

We have turned the task of figuring out what the causes of motion are over to a bunch of mindless nitwits who insist that because the cause is not apparent, it doesn't exist.

"We" has to be separated from "them."

There is the "we" that believes in occult bullcrap answers to motion, explanations that have no meaning, gravity is a property of mass, rotation is the result of angular momentum being conserved in closed systems, orbiting is the result of force imparted from the primordial swirling mass of gas.

That we is them!

And there is the "we" that wants to know why things move, that demands an explanation for the obvious realities in our existence, why objects drop, why the sun comes up, what make the moon move around the Earth, the "we" that is not going to put up with nonsense any longer, that is going to demand a little work from the people we pay to explore these questions as evidenced by a little success in actually answering the questions.

That we is us!

WHY DO WE WANT TO KNOW?

Not knowing the answers to the obvious facts in our reality is probably okay so long as we are not aware of our own ignorance. Thinking that we know the answer to questions of motion because we have made up a bunch of nonanswers and made them real by universal agreement, leaves us dumb and happy. Being able to live with absurdities such as the conclusion that it takes force to move matter on the surface of the Earth, but it takes force to keep matter from moving in straight lines in space is fine if we just want to wallow through life in our own ignorance.

But ignorance has consequences!

All evolved organisms attempt to extend their range of survivability in the environment simply as a result of the drive to continue to exist. Complexity furthers this goal. Moss is rooted in place and come the winter dies in place. Trees can't run away from the forest fire like more complex ambulatory animate matter can. Sentient animate matter can do more than flee the changing conditions of the environment, it can alter the environment itself.

We, in the sense of the "we" that is sentient rather than the insentient "them" that makes up laws governing reality and then forever limits reality to the operation of those laws, we have the ability to produce a picture of the environment when the environment doesn't exist, and it is this ability that allows us to change the pictures of the reality we form to produce a reality that would be more conducive to our survivability and then to go out into the environment and conform the environment to that made-up picture.

If we limit the pictures we form of reality to pictures we have made up, instead of the actual facts that make up physical reality, then the pictures we form to alter that reality are not going to have anything to do with that reality, and the technology we produce to manipulate our existence in that reality is also going to be distorted.

We are going to be washed up a wreak on foreign shores

If we make up a gravity that cannot be manipulated because it is a property of the matter that has to be used to manipulate the gravity, we will never attempt to manipulate gravity, but rather build machines that have to oppose it, planes with wings and engines, rockets with thunder and fire. Instead of attacking gravity directly, we use the atmosphere like fish use water and we do so being as blissfully unaware of reality as the fish are unaware of the water in which they exist.

If we have an inaccurate picture of the physical reality that we occupy, then the technological solutions we create to further our range of survivability within that environment will not accomplish their goal, and we will fail.

The result of the failure to extend our range of survivability is quite simple.

We cease to exist!

It is generally believed that the greatness of Galileo, Kepler and Newton was that they went beyond mere description to discover the "true system of nature" and in the process extracted from flawed and confusing data the simple fundamental laws that control our existence in the universe.

A wonderful achievement, if the goal is to describe something other than reality, which unfortunately is the movement of matter by current forces which, because those forces are constantly changing as a result of the movement of the matter itself, does not conform to our fundamental laws.

Galileo, Kepler and Newton lived when the storehouse of knowledge was extremely limited. Working on the basis of limited knowledge, they claimed to discover universal principles, frictionless motion continues forever, planets sweep out equal areas in equal times, the inverse square law measurable in falling objects controls the movement of matter in space, none of which describe reality.

We know that none of these principles applies to reality simply by observing and measuring reality, by the simple weight of evidence.

And because we know the weight of evidence, we also know the facts. We are not dead men who knew nothing, and therefore had to rely on making up laws and principles to cover our ignorance, we are living people with minds that can be used to analyze physical reality and come up with explanations that provide a basis for an effective technology that can be used to extend our range of survivability into the future.

We know that the movement of the planets are actually chaos, that there is no order in the solar system. This doesn't reconfirm our belief in orderly motion, it tells us that whatever is causing the motion in the solar system is a current force, and that the current force causing the motion, while continually changing, is actually holding the solar system together against the consequences of that changing force.

If from that starting proposition we cannot sit down and come up with an explanation for the forces that cause the current motion in our existence, then perhaps we deserve to go missing, become extinct.

At least we will die happy.

But for crying out loud, we'll die without even knowing why the sun comes up everyday!

5. TURNING BELIEF INTO FACT

DEALING WITH A DEAD-END SCIENCE

Science has been able to foist off on the public a belief that it is responsible for the technological advances that have come from simple trial and error, the testing of physical reality, by engineers whose goal in life is to make things work in reality rather than to explain why things work in reality.

Whenever the efficacy of science is questioned, its practitioners point to things like airplanes and radar, television and elevators, ocean liners and computers to justify science's existence when all of these items are the product of engineering, the trial and error exploration of the limits of reality by practical people attempting to accomplish something in that reality.

Empirical Science finds itself constantly having to justify its existence by engineering accomplishments because the picture it has constructed of the universe is so self-contradictory, so very nonexplanatory, so downright ditsy, that it defies the very reason for which pictures are designed, to provide comprehension. Science has produced a modern artist's universe with made-up absurdities splashed willy-nilly on a canvass with neither dimension nor texture.

But worse from the engineering standpoint, from the view of those attempting to bend the boundaries of physical reality to the advantage of all, science constructs a universe out of theory and then claims that theory to be fact, frustrating attempts to deal with physical reality, in fact, making the actual nature of that physical reality invisible to the hopeful gaze of innovation.

Because Empirical Science claims to measure that which can not be measured, what Bacon referred to as the secret motions of things, it couches its conclusions about physical reality in mathematical formulae which take the expression of one things equals another thing.

Because these mathematical expressions cannot handle an

unlimited number of variables, and because the universe is composed of limitless variables, to claim that mathematics can provide answers to the unknown facets of physical reality, the secret motions of things, mathematicians have to limit the terms of their equations by creating laws which do not vary in order that the variables the laws describe may be eliminated from consideration.

As a result, mathematicians, in producing a science which is supposed to describe physical reality, come up with invariables such as matter can neither be created nor destroyed, gravity is proportional to and therefore a property of matter or mass, motion is conserved in closed systems, objects always move in a straight line unless a force acts on them to change that motion, or with energy and mass interchangeable in a formula that relates them to the speed of light, no energy can cause mass to exceed the speed of light.

All of these negatives are utter nonsense, have no basis in reality, violate the basic tenet of Empirical Science that statements be testable and do not add to the task of coming up with a consistent picture that allows us to comprehend the universe.

But from the point of view of the practical engineer, the negative aspect of the statements, the negativism coming from the need to make the physical reality the statements deal with static, invariable, nonoperative, cause the statements to block the possible and close off attempts at trial and error experiments which might advance our ability to physically manipulate our environment.

The simple statement that gravity is a property of matter precludes any attempt to manipulate gravity because properties are merely inherent in matter and as a result, simply exist.

Without the ability to envision the possible, the engineer expends energy probing the impossible that is left, how to overcome the gravity, how to build machines and structures that can be pushed against gravity rather than how to build machines and structures that can utilize the force that makes objects fall to themselves move.

THE DIFFICULTY IN PROVING THE FACT THAT THE EARTH ORBITED THE SUN LED TO THE CONFUSION OF FACT WITH CONCEPT

It was not the Copernican realization that the Earth moved around the sun that defined modern Empirical Science. It was the inability to demonstrate it that doomed the search for a clear picture of how the universe operates to a four hundred plus year journey that, in a process similar to the massive mistake the Copernican realization corrected, produced clones of the mistake so that the errors in the tenets of Empirical Science are as numerous as the tenets of Empirical Science itself.

Copernicus made a simple statement. He said that the sun and other planets don't move around the Earth, rather that the Earth and the other planets move around the sun. Aristarchus had made the same statement seventeen centuries earlier, but the one hundred years of navigational experience that preceded the Copernican statement lent credence to the statement.

Credence or no, the statement was not provable. No one could take a ruler, walk out the door and measure the Earth going around the sun.

Some facts in reality are measurable, others aren't. Mathematics can, at times, be used to measure that which can't directly be measured. We can easily measure the distance from the front door to the street. We can use the relationship of lines and angles to compute unknown measurements from measurements that we know, measuring the height of the electric pole at the end of the sidewalk using the angle of the sun and the length of the shadow the pole produces. Processes can even be devised that allow us to measure the distance to the moon. We can, now that we know the relationship between the planets and the sun, devise mathematical procedures such as determining solar system distances by parallax that could only exist if the Earth traveled around the sun and for that reason prove that it does so.

But what goes around what is a relationship and relationships are both factual and conceptual. The bare distance between two points is a fact with no conceptual content. Facts in and of themselves neither convey meaning nor require understanding. The changing relationship between two objects, however, produces a fact that requires visualization, and therefore conceptualization. The changing relationship between two objects is time, which is a concept, and the relationship of what is where also requires conceptualization.

The Ptolemaists had no difficulty handling the simple measurable fact of static distance, and they had no problem with conceptualizing changing relationships, they, after all, could observe the sun and the moon move around the Earth.

The Copernicans were equally adept at handling both the simple facts and the conceptualizations of moving objects.

What neither could handle was the relationship of what went around what, and while this is a simple fact, when the simple fact is not known, then how the facts are viewed is bald conceptualization.

WHAT WENT AROUND WHAT WAS A FACT BUT, BECAUSE THE FACT WAS UNKNOWN, ATTEMPTS TO PROVE THE FACT WERE ACTUALLY ATTEMPTS TO REFUTE A MISCONCEPTION AND THUS REQUIRED CONCEPTUALIZATION

The science of the Ptolemaic solar system, the Aristotelian concept of the universe, was called Philosophy and the practitioners of Philosophy were called Philosophers, quite the opposite of the meaning today, where anything philosophic is the opposite of scientific. Because the Philosophers believed that the sun went around the Earth when in fact the Earth went around the sun, they held a misconception about the factual relationship between the Earth and the sun.

On the one side there was a bad concept, a concept that opposed the facts, and on the other side there was a good concept, a concept that agreed with the facts.

But facts are not concepts.

Facts are facts!

Once we had the correct picture, we didn't need conceptualizations to prove it, we merely needed to form the correct factual picture. Facts merely need demonstration. Of course, knowing that the Earth and the planets move around the sun raises all sorts of other questions about the facts, questions like what causes the planets to move, that need conceptualization but the relationship of the planets to the sun no longer provides a basis for conceptual arguments.

When Galileo went up against the Philosophers in an attempt to prove the correctness of Copernicus, he was in the middle of a conceptual argument as to what the facts were. He was saying, "Mr. Philosopher, your concept of the solar system is wrong, it is not fact. The picture Copernicus painted of the solar system is correct." And although Galileo unfortunately never reached the point that he could finish the statement, if he could have conclusively demonstrated to the Philosophers that the Earth went around the sun, then he could have said, "What we had to argue as concepts because we didn't have the facts, the relationship of the Earth to the sun, is now a fact and no longer requires any argument. Facts don't have to be proven, they merely have to be demonstrated."

However, because Galileo couldn't demonstrate the fact, he had to diddle around with ways to prove the fact, and because it wasn't a demonstrated fact, the proof was an attempt to prove a concept.

The science that has grown up since Galileo has, as a result, confused the two, fact and concept.

Because the relationship of the matter in the solar system was a fact, it could be demonstrated. However, showing a concept about one set of facts to be correct does not mean that all concepts can be shown to be facts!

GALILEO INVENTS A PROCEDURE DESIGNED TO PROVE CONCEPT TO BE FACT

Simply demonstrating the possibility of the solar system by viewing the Jovian system did not satisfy the proofs needed by the Philosophers who were used to mathematical demonstrations of relationships and logical reasoning to express how those mathematical demonstrations applied to the facts.

Galileo, undertaking the thankless task of providing the Philosophers proof on their own terms, sought to center his formal proof that the Earth went around the sun on the demonstration that the Earth rotated on its axis thereby giving the sun the appearance of movement. However, he had no explanation for the Earth's rotation because the force that maintains the Earth's rotation is not apparent. He therefore set out to show that the Earth, set in motion at the creation of the universe, could continue to rotate because its rotation in frictionless space was unopposed. Once he had shown that the Earth could rotate forever in frictionless space, he planned to demonstrate the rotation by showing that the tides moved as a result of that rotation.

Galileo named the force that maintained his perpetual rotation inertia. To prove that inertia existed, he rolled balls down an inclined plane onto a level surface. He reasoned that the only thing that would slow the balls down on the level surface was the friction between the surface and the ball and that if he had a frictionless track that circled the globe, the balls would roll around that track forever.

These unmeasurable fantasies could be dismissed as harmless if they weren't at the center of Empirical Science, "empirical" meaning measurable!

Because Galileo couldn't devise a frictionless circular path around the Earth to prove his concept of inertia to be fact, he had to come up with some other way to demonstrate that inertia was, in fact, a fact.

Half a century before Galileo was carrying on all of these ruminations, a practical trial and error advocate named Tartaglia had gone out onto the proving grounds and shot off cannonballs at different angles in order to come up with the angle that would cause the cannonball to travel the furthest distance. After much banging around, he determined that the optimum angle was forty-five degrees.

Galileo allowed the balls that had rolled down the inclined plane to roll off the end of his workbench and measured the same angle, at forty-five degrees, the inclined plane produced the maximum trajectory.

He then said he had, in the confines of his study, by mathematics alone, proven that which was proven by trial and error on the proving grounds. It was acceptable procedure, then, to use mathematics to prove facts that were not capable of being directly measured.

Of course, Tartaglia had actually shot real cannonballs out of real cannons to produce real measurements. By duplicating the facts in his study, Galileo had not proven a fact that was not capable of direct measurement, he had just measured the same fact in a different way, measuring real balls that had rolled down a real inclined plane rolling off the end of a real workbench.

He hadn't proven anything let alone a concept, inertia, to be a fact.

Concepts about factual relationships can be demonstrated by measurement to be fact.

Concepts, however, can never be proven to be facts!

DISTINGUISHING BETWEEN CONCEPTS ABOUT FACTUAL REALITIES AND CONCEPTS THAT HAVE NO BASIS IN FACTUAL REALITY

The sun exists and the Earth exists, and the two move with respect to each other. Even if we don't know which is going around which, we know that one is going around the other. We are dealing with facts, and the relationship of those facts will eventually out.

Inertia as a force that can cause something to move, on the other hand, is simply a made-up fantasy. The Earth spins on its axis, that's a fact. What is making it spin on its axis is another kettle of fish altogether.

When Tartaglia shot his cannonballs off on the proving grounds, the force imparted to the cannonballs by the charge in the cannon was the existing force that propelled the cannonballs against the force of attraction that would otherwise cause the cannonballs to seek their lowest point, the barrel of the cannon. The existing force that put the cannonballs in motion no longer existed, so it was left to the Earth's attractive force to overcome the force that was imparted to the cannonballs by the charge.

The force of the charge did not in some occult manner take up residence in the cannonball. The cannonball was just a dumb piece of matter. Once it was set in motion by the charge, where it would land would depend on a number of external factors acting upon its progress, factors that included the air resistance, Galileo's friction, and the rotation of the Earth, the Coriolis effect, even the wind, but the primary force that would affect the course of the cannonball would be the force of attraction produced by the Earth. As it happens, when the angle of the cannon is forty-five degrees, the time it takes the Earth's attractive force to degrade the force imparted to the cannonball is at its maximum.

When Galileo rolled the balls down the inclined plane, he duplicated the forty-five degree angle of Tartaglia's trial and error cannon. When the balls rolled off the workbench with the force imparted to them from the plane inclined at a forty-five degree angle, the time it took for the Earth's attractive force to degrade the force imparted to the balls by the inclined plane was at its maximum.

In neither case does the matter that makes up the objects obtain anything other than motion that resulted from an existing force. Objects don't move with respect to historical forces, they come to rest with respect to existing forces, so that both

Tartaglia's cannonballs and the balls from Galileo's inclined plane were set in motion by a current force and after the current force was removed, the balls began to come to rest with respect to existing forces.

To say that a force exists that causes the balls to continue in motion after the force that set them in motion ceases and call that force inertia is to create a concept to describe the motion that occurs between the time that an object is set in motion and the time it comes to rest with respect to the current forces acting upon it.

With no current force moving the ball during this period, the term inertia to describe that absence of current force is a meaningless concept, a concept that applies to a fact that does not exist in reality.

To then say that inertia causes a planet to rotate is to extend a nonfactual concept to the cosmic level.

When we are arguing about what goes around what, we are arguing about what the facts might be.

But we are still arguing facts!

When we are arguing about what makes objects move, we are arguing about intangibles that might produce the facts we see. We have no way of knowing precisely what those intangibles might be because they don't present themselves as tangible matter to which we can apply our tangible measuring sticks, or even our intangible mathematical measuring processes.

BACON RECOGNIZED CLEARLY THE DIFFERENCE BETWEEN FACTS THAT CAN BE MEASURED AND CONCEPTS ABOUT FACTS THAT CAN NEVER BE MEASURED

Sir Francis Bacon recognized that it was necessary to clearly delineate the line between facts that could be measured and facts that could never be measured.

Galileo had used deductive reasoning to conclude that the concept of inertia was a fact.

There is in reality fact and concept, facts being measurable and concepts being what we make up in our minds to make sense out of the facts. Empirical Science claims that by verifying its concepts, which it calls hypotheses, by finding facts those concepts deductively predict, it can protect itself from unfounded concepts such as the Ptolemaic mistake that the sun went around the Earth. Induction, on the other hand, is the process of producing concepts to explain facts, and inductionists claim that because the concepts precisely describe the facts, they too are factual.

In fact, the inductionist's concepts are the Empiricist's hypotheses and both end up in the same place, with concepts that are mistaken as fact.

By blurring induction and deduction, Modern Science creates a sort of moving target behind a claim of consistency, the scientific method, that avoids directs hits. When a direct hit looms, Modern Science claims that all of its conclusions are simply falsifiable, shifting the argument to epistemological grounds by claiming that surviving falsification produces consilience, agreement with facts, and therefore facts. If all else fails, Modern Science can shift the argument to religious territory by design arguments, thereby hiding God's hand in nature, the swirling mass of gas and big bang being the most prevalent.

In actuality, the mind works by storing information and recalling that information in the form of pictures. When a picture has a missing element, we look for the missing element so that we can complete the picture. If we are dealing with measurable facts, filling in the missing element, finding the missing fact creates a bigger picture and the new picture will now have holes in it, more missing elements, for instance, what makes the Earth move when we find it moves around the sun. This is the trial and error nature of the scientific method that produces technology. We recognize a missing fact, then go out and find it, measure it, size it, and put it into the picture we already have.

However, some facts are missing that are clearly there but appear to be missing simply because they aren't measurable. Objects fall but we can't pick up and measure what is making them fall. We see movement in the heavens, but without apparent force doing the movement, out telescopes don't produce a picture of what is causing the movement.

We have to come up with some sort of explanation and we do, and so long as the explanation doesn't produce a picture with more missing pieces in it, we never question the explanation.

This is what we have done with the secret motions of things. Bacon said we have to continue to perfect concepts about the hidden motions of things by uncovering more and more facts, that our concepts about the facts can never be facts, but have to evolve as our knowledge about the facts evolves.

Galileo had done quite the opposite, starting out with Tartaglia's physical measurements and reasoning that duplicating those measurements in his study would show that there was a force that caused the movement of the balls when there was no force present to move them. He then deduced that the friction acting on the balls degraded the balls movement and that without that friction, the balls would move forever, a purely inductive statement that, because it was derived from deductive reasoning, he took as fact.

Bacon said that the cause of movement, rather than the measurement of the movement itself, was secret. The rate at which an object falls toward the Earth is a physical measurement which results in a fact, but what makes the object fall is a secret. We can't, he said, measure the nature of the force that causes objects to fall or planets and moons to rotate and orbit because it is not apparent, it is not physically capable of measurement.

He labeled measurable facts to be facts, physical realities established by measurement and verification, and he labeled what causes movement to be the secret motions of things, and then stated that what caused the secret motions of things could neither be measured nor verified.

The solar system was too big to allow us to measure it directly, but because it was composed of the movement of matter, we could produce measurements as to its size and motion and the relationships of the matter that produced the motion.

What caused the motion, on the other hand, was a secret that was consigned to areas that were too small for perception.

It was necessary, however, to search out the nature of the secret motions of things, and it was to provide a basis for this search that Bacon came up with the process of inductive thinking.

Bacon said that the secret motions of things cannot be proven as fact, it can only be induced from the facts and anything that is induced from the facts is dependent on the facts and is not itself a fact. Because we cannot induce the secret causes of motion, we obviously do not have enough facts. Thus, what we have to do is go out and gather as many facts as we can, and eventually, when we know all of the facts there are to know, then we will easily be able to induce the secret motions of things from those facts. In the meantime, we propose concepts and change those concepts as our knowledge of the facts increases.

Bacon's induction is therefore a generalization of Ockham's razor which states that what explains the most with the least prevails, Bacon demanding that the explanation consistently be revised as our knowledge of the facts expands, not that we take a few facts we know about, generalize laws, facts expressed as ignorance, and forevermore control the facts we see by the laws we have made up.

Induction is a process for coming up with concepts that can be used to explain what we see the facts doing.

It is not a way to come up with facts or laws that can then be used to control the nature of the facts.

LAWS ASSUME AWAY THE SECRET MOTIONS OF THINGS

Galileo's inertia was a historic force that explained away current motion. When Galileo tried to use the tides to demonstrate the Earth's rotation, he found that the tides would create a friction, a current force, that would degrade the Earth's inertia, a historic force, causing the Earth to stop rotating.

This debacle should have brought to an end occult attempts to explain away current force, Bacon's secret motions of things, by saying they became a property of the thing doing the moving.

Instead, it just altered Galileo's word "inertia" to mean a resistance to a change in motion rather than a cause of the motion itself.

The cause of motion became momentum, which was precisely what Galileo had thought of as inertia. Instead of the Earth rotating because of inertial forces set in motion when God created the universe, it rotated as a result of the angular momentum given the matter when God created the universe.

Neither, of course, was a cause of motion, rather merely dumb explanations for continued motion!

Angular momentum was subsumed in Newton's laws of motion, one of which stated that an object will continue to move in a straight line unless a force acts on it to change that motion, its inertia being to resist anything but straight-line motion.

Same occult bullcrap that Galileo had made up to explain the Earth's rotation!

Some force is applied to an object at sometime in the past, that force becomes a part of the object, is a property of the object, seeps into the object's pours, and then the object forevermore contains that force.

There is a big difference, however, between inertia and momentum, a difference that Galileo had recognized when he visualized the balls rolling around a frictionless course forever but which he was able to ignore when he was dealing with the rotation of the Earth. When he applied his frictionless motion to rotation, he didn't have to deal with the fact that the primary force that degraded the force imparted to Tartaglia's cannonballs was the Earth's force of attraction.

It was the Earth itself that was rotating.

With Newton's laws of motion, when the force of attraction caused the object traveling in a straight line to change its motion to circular motion, it did not degrade the motion inherent in the object, the object's momentum. Newton lifted Tartaglia's cannonballs off the surface of the Earth into frictionless space so that the force of the Earth's attraction equaled their straight-line momentum. Newton said that like Tartaglia's cannonballs in Galileo's perpetual circular courses, the Earth's force of gravity balances the straight-line motion of the moon.

With the two forces balancing exactly, neither was used up.

Gravity as a force that diminishes when it doesn't do anything but doesn't diminish when it does, when it alters the straight-line motion of planets into circular motion, and momentum, which isn't degraded like the momentum of objects on Earth is degraded by the Earth's attractive force, are fantasies at their most extreme. They are lies writ so large they can be naught but fact.

In reality, they're just silliness, a display of ignorance so great that only fools would attempt to defend them.

As a result, so that the holders of these beliefs can escape the label of fools, the concepts have to be rendered unnecessary of defense. To do this, they have to be expressed as natural laws of the universe, things that simply exist. Objects travel in a straight line unless a force acts upon them to change that motion, angular momentum is conserved in closed systems, the inverse square law of falling objects applies to planetary orbiting, these are not physical facts that have to be proven.

They are universal laws which the genius of man has induced from the facts!

In other words, we are too dumb and lazy to distinguish between fact and fiction, between reality and our made-up concepts of reality. We'll just call what we make up law and this will allow us to believe that we know the secret motions of things.

By creating laws that assume historical forces for current motion, we simply make the existing force that produces current motion invisible, no analysis need be applied, because we now believe we know everything there is to know about the secret motions of things.

And we know absolutely nothing!

BELIEVING THAT NEWTON ACTUALLY PROVED A CONCEPT TO BE A FACT, THAT A PROPERTY OF SOMETHING COULD ACTUALLY CAUSE SOMETHING ELSE TO PHYSICALLY MOVE, WHEWELL LOOKED TO BACON FOR A BASIS TO EXPLAIN NEWTON'S ACCOMPLISHMENT

There is nothing factual about Newton's mass/gravity fantasy. Newton relied on God for his straight-line motion, we rely on a swirling mass of gas which if made up of mass would clump instead of swirl, straight-line motion doesn't exist in the universe, objects don't move unless something causes them to change motion, objects come to rest with respect to the forces acting upon them, gravity, a straight-line force, provides no basis for circular motion in the solar system, the inverse square nature of Kepler's laws upon which Newton based his idea that the planets fall toward the sun the same way that a rock falls to the surface of the Earth are disprovable by any first grader with a stick, the list of absurdities is endless. The one thing that the mass/gravity concept can do is compute, using the inverse square law and the law of straight-line motion, what's in the center of planets and at various points around the universe where there is no visible matter.

And these computations, which can never be checked, provide useless information.

But mass/gravity is as real as the effect the moon has on the tides. The fact that the friction of the tides would have stopped the rotation of the Earth in frictionless space had long puzzled Philosophers and later Naturalists, the predecessors of Empiricists, so that when it was found that the moon could exert a gravitational pull on the water, literally lifting it out of its seabed, sighs of relief were heard round the world as the friction problem was solved.

No tides can be predicted by the movement of the moon, just as no evidence can be produced to show that the attractive force is a property of physical matter, but that is of no concern when everybody on the planet believes it to be so.

And it was that belief that Whewell wrestled with when he tried, a hundred years after the fact, to relate Newton's discovery with the procedure he had to have used to come up with whatever Newton discovered so that he could show that there actually was a connection between the attractive force and physical matter.

It is hard to show how a fact was discovered when a fact wasn't discovered. When Columbus discovered America, it was simple to show the winds he took to get there and the winds he took to get back by getting on a ship and going there and coming back.

When facts are facts, facts are easy to demonstrate. Even the fact that the Earth goes around the sun is easy to demonstrate now that it is known to be a fact. Erroneous concepts about facts can be corrected by measurement.

But when concepts are mistaken for facts, then its not so easy to demonstrate that they are facts.

Newton's mass/gravity fantasy is more than a concept, it is a group of interrelated concepts that has no relationship to reality, so demonstrating its factual nature is not a simple thing to do.

And yet Whewell knew that Newton had proven a concept to be a fact, so he knew that he should be able to sit down and demonstrate exactly how he had done it.

And he demonstrated and demonstrated as chapters became books and books, volumes and volumes, sets.

If facts are easy to state, why then did it take Whewell volumes?

Because he wasn't dealing with facts. Every time he got to the core of the mass/gravity matter, every time he faced the straight-line law of motion, he was in a conceptual minefield because he knew he wasn't dealing with a fact. But because it is true that an object will always move in a straight line unless a force acts on it to change that motion, after all, who can disagree with the statement that an inch tall elephant will always fit in a watch pocket, Whewell was faced with the question of where Newton came up with it. It was not deducible from anything, it was just something that had to be operating in the universe if gravity were a property of matter and as gravity was a property of matter, as elephants are an inch tall, then it was something that was operating in the universe.

It's either deduction or induction. If a true statement can't be shown to be deduced, it must have been induced. Aristotle had talked about inductive thinking but everything Aristotle talked about was wrong because he was wrong about the relationship of the Earth to the sun. Bacon had also talked about inductive thinking and because Bacon knew the correct relationship of the

Earth to the sun, Newton must have been using the Baconian inductive process to derive his true statements about motion.

Looking at the orbiting planets, and the fact that they could be going in a straight line if they weren't falling toward the sun, Whewell had to conclude that the only way that Newton could have derived his fact, his law of motion, was by induction.

Bacon said that the secret motions of things would be obtained by allowing facts to produce inductive thoughts that explained the facts, and that's precisely what Newton had done.

Thus, Newton used Baconian induction to come up with mass/gravity!

MISTAKING FACT FOR UNDERSTANDING

The secret motions of things requires the use of the mind to understand. Because we cannot form a picture in our minds of the forces at work moving matter in the solar system, we have to produce concepts that explain that motion.

Assuming away motion is not a process that produces concepts that explain motion. By manufacturing the fantasy of a swirling mass of gas to explain planetary orbiting and rotation, we have assumed away the cause of planetary motion.

Empirical Science justifies this by its requirement that facts be verified. If a Naturalist goes out into the field and discovers a giant bone, he will be automatically accused of fraud. When a dozen Naturalists follow him and they all find identical giant bones, the giant bones will be accepted as fact and concepts will be created to explain their existence.

However, when the things sought aren't retrievable, then Empirical Science has no way to deal with them. Thus, when someone spots Nessie in Lock Ness, or Bigfoot in the great northwest, when reports of unconventional objects moving in the skies abound, when alterations in the environment such as crop circles appear without a physical basis for their existence, Empirical Science has no recognition capability to deal with the phenomena. It labels them frauds or worse!

If it can't be put on a shelf in a fifty museums, then it isn't.

Empirical Science approaches the secret motions of things in the same way. There's no apparent force causing current motion in the solar system. Even though on Earth any movement requires the application of current force, because there is no current force apparent in the solar system to cause the current movement, Empirical Science simply says it doesn't exist, that the movement of matter in space doesn't require current force like it does on Earth.

Because the matter in the solar system is doing something, it is moving, measuring that movement is used as a substitute for understanding what is causing that movement. The Newtonian explanations for that movement, straight-line motion and gravity, are bald concepts with no basis in reality other than the fact that gravity obviously exists, it causes us to stay glued to the surface of the Earth and released objects to drop.

It is patently absurd, however, to say that what causes objects to drop makes them orbit.

Whewell said that the complimentary use of induced laws allowed Newton to come up with an explanation for all of the motion in the universe when Newton didn't come up with an explanation for anything, let alone motion. Whewell pointed to Bacon as justification for Newton's claimed achievement when Bacon had said that facts had to be verified, but that the explanation for facts could only be conceptually induced from the facts.

Whewell called Newton's fantasies of straight-line motion and universal gravitation as embodied in the inverse square measurement applied over infinite space laws that Newton induced which then allowed him to explain motion, but Bacon would have called them concepts that were made up to explain a series of facts and were only as good as the facts that they explained.

Bacon's induction was a living process in which the explanations, the concepts induced from the facts, were only as good as the facts they explained and the whole purpose of seeking out and verifying additional facts was to alter, to expand the concepts induced from the available facts.

Bacon didn't say, make up a bunch of bullshit from a purposely limited set of facts and then say, because the bullshit explains the specific set of facts it was made up to explain, it is now a fact to be treated with greater respect than actual facts in reality are treated.

Bacon said induction required that concepts explain verified facts. Newton made up a bunch of concepts that explained each other, the moon would be falling if it weren't going in a straight-line which it isn't because it is falling, and Whewell comes along and says Newton's right because he used Baconian induction.

Oh, the injustice!

WHAT'S REALLY GOING ON HERE?

J. F. W. Hershel, the grandson of the Hershel that is renowned for discovering Uranus, perhaps in an attempt to explain why, if his grandfather's computations were correct, he had Uranus going in the opposite direction it's actually going, found himself constrained to clarify Whewell's use of Baconian induction to justify the Newtonian mass of conceptualizations. By this time, it was clear that Newton's procedure failed to demonstrate that gravity was a property of the matter around which other matter orbited. Newton had used volume to determine the amount of matter in the moon and the Earth, and thus compute the amount of force that each was exerting on the other so that he could mathematically demonstrate that the force was related to the amount of matter.

Computing the force on the basis of volume didn't work for the Earth and the moon, and it certainly didn't work for anything else in the solar system, so the whole pile of Newtonian bullcrap was simply nonoperative.

And yet, everybody knew that Newton had come up with the right conclusion, that the planets operated in accordance with the same inverse square law that caused objects to drop on Earth.

Instead of using matter to predict the force, it was concluded that, because it was measurable on Earth that objects fall in accordance with an inverse square measurement, that measurement was actually a law, it didn't need verification, proof, and that the amount of matter in the planets could therefore simply be computed by applying the inverse square law to motion.

Hershel looked at Baconian induction and said that the concept induced had the greatest validity where it was in agreement with all of the facts but one, and then that fact was found. In this view, Newton had induced a process that was in agreement with all facts, the movement of the moons around the planets and the movement of the planets around the sun. The only thing that the concept didn't agree with was the position of the planets. However, because Newton's induced concepts could be used to predict the position of any planet at any time, the missing fact was continually being proven and Newtonian Mechanics provided a self-generating proof of its own validity.

That's why the claim is made that the Newtonian scheme is the most proven theory in history.

Strip Newtonian Mechanics away, however, and we find the same process we have if we want to find out the timing of the tides, we show up and measure them. What has happened in the past will continue to happen in the future. The planets were here last year, they'll be hear next year.

And interpreting Newtonian Mechanics as producing a self-generating proof of its own validity, and relying on Whewell's persistently lengthy claim that Newton derived his procedure from Baconian induction, Baconian induction becomes the production of simple statements of fact that agree with measurable phenomena, the planets and the moons would be moving in a straight-line but for the fact that they are falling toward the sun or the planets which they aren't because they are moving in a straight line.

Induction becomes computer modeling where piles of ad hoc bullcrap concepts are tweaked to adjust the model to observed reality, and in the process, turned into facts.

This is called enumerative induction today, but currently generalized as induction by Empiricists. Bacon characterized it as nothing but static description, requiring his concept of induction to be capable of growing with the facts.

The reason everyone has trouble applying Baconian induction is that Baconian induction involves the generation of concepts to explain facts, concepts are only as good as the facts they explain, and concepts can never be facts.

The reason everyone has trouble applying Baconian induction, for crying out loud, is that Baconian induction requires the use of the mind!

However, the evolution of Galileo's belief that he could, by mathematics alone, in the confines of his study, demonstrate things which he could not measure, his belief that a ball would travel around the Earth forever in a frictionless path, to be facts into Hershel's belief that a concept derived from a number of facts that explained a missing fact became a fact when the missing fact was found, all in the name of Bacon, thanks to Whewell's belief that Newton had turned a concept, that gravity is a property of matter, into a fact using Baconian induction has formed our picture of physical reality with a mindless process that produces a picture of physical reality that has no basis in fact.

Today, we say we isolate a couple of facts, bad enough in itself, derive a testable hypothesis from those facts, testable meaning that the facts deductively predict an unknown fact that can be verified, or in the current vogue, a hypothesis that can be falsified, and then test for the fact or attempt to falsify the hypothesis. If the predicted fact is found, which it generally will be sooner or later, after all, that's what we were looking for in the first place, then the hypothesis itself becomes fact. If the hypothesis escapes falsification, if its stands the test of time because it is so fanciful that no test in reality can dislodge it, the mass/gravity fantasy being the preeminent example, it becomes fact.

What we end up with are a bunch of nutty propositions accepted as fact because they predicted a fact that was eventually found to be true (facts like quarks, black holes, dark matter, the sort of thing that can never be falsified), and a bunch of nonsensical notions like event horizons for black holes, the pervasive big bang explanation for the formation of matter in space to compliment the fact that when matter is exploded on Earth it comes apart, or even the venerable swirling mass of gas.

Empirical Science, by its refusal to recognize that it takes concepts to understand facts, the Baconian inductive process designed to grow concepts until they approximate reality, and instead believing induction to be a process that proves concepts to be fact by making them law, has conditioned itself to accept not just mistakes about physical reality like the sun travels around the Earth, but any absurd statement that happens along.

Facts never have to be proven. If they are, they exist, they are measurable, they are facts.

Concepts can never be proven to be facts.

Concepts can never be proven to be facts.

Concepts can never be proven to be facts.

Spin around three times and say it again.

There is no process, no procedure, no hocus-pocus bullshit legerdemain that will turn a concept into a fact.

There are things that we will never know for certain.

However, if we make up things and then believe them to be fact, there are certain things we will never know at all!

Mass/gravity is a concept, not a fact. Angular momentum conserved in closed systems is a concept, not a fact. The swirling mass of gas is a concept, not a fact. Inertia or momentum, or any combination of either is a concept, not a fact. Deduction, induction, falsification, indodeduction are all concepts, not processes that produce facts. All objects in space traveling in straight lines but for the fact that they aren't isn't even a concept, its simply ridiculous, and to claim that current motion is the product of historic force is to proclaim massive ignorance.

If we want to figure out the current forces in the universe that cause matter to move then we are going to have to recognize that there are current forces in the universe that cause matter to move!

6. WE CAN SEE LIGHT FROM STARS AT THE END OF THE UNIVERSE AND THE BEGINNING OF TIME AND FIBBER MCGEE AND MOLLY ARE OUT THERE SOMEWHERE

A MEASURABLE PHYSICAL REALITY

What happens when we light a match?

The match stem ignites, begins to burn and the match begins to emit light.

What physical presence does the light have?

The light has a physical presence in every direction that is not blocked by matter such as the match stem itself.

What geometric shape is produced when something has a physical presence in all directions?

When something has a physical presence in all directions, it forms a sphere. A sphere is a three-dimensional surface all points of which are equidistant from a fixed point.

Doesn't light fill the sphere?

Light produces something more than just a static sphere because the light is moving at approximately one hundred and eighty-six thousand miles a second. At each instant the match is burning, it is emitting a sphere that is expanding like an inflating balloon out and away from the surface of the match at the speed of light. And because the match is burning over a period of time and is emitting a new expanding sphere each instant it is burning, the light that is being emitted from a burning match is a series of expanding spheres traveling at one hundred and eighty-six thousand miles per second away from the surface of the burning match.

What do we know about spheres?

We can determine various measurements relating to spheres by reference to specific mathematical formulas. We can determine a sphere's radius, its volume, and most importantly, we can determine the area of its surface.

What is the formula for computing the area of the surface of a sphere, in this case, a sphere of light?

The area of a surface of a sphere of light is $4\pi r^2$ which is four times the constant pi times the radius, the distance to the center of the sphere, squared.

If there are two spheres of light being emitted from a match, how do they compare with one another?

The formula for determining the areas of all spheres with a common center has two constants, four and pi. The only term that differs is the radius, the distance to the center of each sphere. Thus, any sphere created by the surface of an expanding sphere of light being emitted by a match is related to all other spheres in that expanding sphere of light by reference to its radius squared, the square of its distance from the common center.

As a sphere of light from a burning match expands, its surface expands with the square of its distance from the match.

What does the surface area of an expanding sphere of light tell us about the light itself?

The surface area of any sphere in an expanding sphere of light is the surface area over which the light is distributed. All of the light that was emitted at any instant the match was burning is distributed over the surface of the expanding sphere, and as the sphere expands, its area becomes larger with the square of its distance from the match. The light is, therefore, being increasingly distributed over an expanding surface area, an area that is expanding in proportion to the square of the distance from the center of the sphere.

What is the effect on the presence of the light on a given surface area as the surface area of the expanding sphere becomes larger?

The larger the area the light has available for expansion, the less light there is available for any given area in the expanding sphere of light. As an expanding sphere of light grows larger, the same amount of light, distributed uniformly over that surface, becomes less for any given area of the surface.

What is the relationship, then, of the light that makes up an expanding sphere of light?

With the area of the sphere expanding with the square of its distance from the match, the light diminishes inversely with the square of the distance over which it expands. As the distance over which the light expands becomes larger, the amount of light becomes less and the amount that the light becomes less is in inverse proportion to the amount the distance from the match becomes larger squared.

This leads to the physically measurable observation that the amount of light at any point from its source is inversely proportional to the square of the distance from that source.

THAT WHICH DIMINISHES UNIFORMLY DISAPPEARS ENTIRELY

We have all been treated to the challenge of figuring out how long it would take to drink a quart of milk if each swallow were exactly one-half the prior swallow. Saying that we would never finish the quart of milk because half of half is always something plays into the hands of those anchored firmly in unreality because such word and number tricks ignore continuity and perspective, the inability to continually take half a swallow and the perspective of massive lips drinking molecules and atoms.

We don't have the foggiest idea what light is, but this shouldn't stop us from concluding that, because it diminishes inversely with the square of the distance over which it moves, distance will eventually eliminate the light and the light will cease to exist.

It doesn't take a knowledge of what light is to realize that with light diminishing inversely as it expands with the square of the distance from its source, it will eventually expand out of existence.

This is not too hard to test, and has, in fact, been tested in reality repeatedly in measurements of candle power over distance. This is not an interpretation of fact or a conclusion about a fact, beliefs such as interference patterns or angular momentum, but rather a simple, measurable fact.

When we are too far away from a source of light to see the light, our failure to see the light results not from a failure of acuity on the part of our eyeball, but on the fact that the light has simply expanded out of existence. We could get some binoculars or a telescope and improve our distance a little, gathering in some of the remaining rays that are too weak to register on our naked eyes, but even the strongest telescope in the world can't see light that isn't there, and light that has expanded out of existence is just that, simply no longer there.

Just as we won't see light from a match when the match has gone out, we won't see light from a match if we are so far away from the match that the light has expanded out of existence before it reaches us.

AN INTERESTING QUESTION

If light expands over the surface of an expanding sphere, diminishing inversely with the square of the distance as it does so, then how is it that empiricists delude themselves into believing that they can see light from stars at the end of the universe and the beginning of time?

The fact that those deluding themselves into believing that they can take relationships, the space between objects and the changing space between moving objects, space and time, and quantify them so that they have an end and a beginning, will easily delude themselves into believing that they can see forever doesn't entirely answer the question.

Delusions need a basis in belief to take hold.

The basis for the delusion that light travels forever is found in the belief that finding a fact predicted by a belief proves the belief because it allows facts to be drafted into the service of beliefs.

CAN'T BLAME THIS ONE ON NEWTON

It's truly a reflection of the ability of the consensus to fashion the shape of the world after its desires that the only assertion Newton made that was close to reality was the only assertion Newton made that the consensus denied.

Newton, because he needed little corpuscles of matter to compute the amount of matter in the Earth and moon by volume to prove that gravity was proportional to, and therefore a property of, matter, thought of everything, including light, as made up of little corpuscles.

Light as a particle, he reasoned, would account for why light bounced off of an obstacle, casting shadows as it did so. Light as a particle didn't, however, explain interference patterns, little ripples of light produced at the edge of the shadow as if the light were going around the obstacle, an effect that was pronounced when the obstacle was something like a pole, where the light had a tendency to move as if it had curled around the obstacle altogether. Interference was more analogous to water waves which could pass a pole unaffected.

More decisive in determining whether light was a particle or a wave was the interesting fact that for light to be a particle, it had to be something, it had to be emitted, and it had to therefore be part of whatever it was that was doing the emitting.

The question of whether light was a particle or a wave was actually a question of whether light was something or nothing!

If light were like a water wave, then it didn't have to be anything because water waves were merely a disturbance of the water and had no independent existence outside the water. The water was merely a medium for the waves produced by the disturbance with the waves merely passing through the water as a temporary disturbance.

To make a determination that light was actually something, that it existed as something physical in physical reality would require finding out just what that physical something was. That was a little too much like work. If Empirical Science can get away with making up some nonsensical term, gravity, for the most important force in our existence, the force that makes objects drop, and delude itself into believing that it understands what that force is by reference to the word, why can't it do the same for another obvious force in our existence, light?

If the choice comes down to explaining how something worked that didn't exist and how something worked that did exist, then it was a no-brainer.

It's a whole lot easier to explain something that doesn't exist!

We couldn't call light water waves, though, because light obviously has no similarity at all to water waves. We don't drown in light unless we're poets, and poets aren't empirical.

But, if we can't call light water waves, we can still say light is like water waves and accomplish the same thing.

PROVING THE WATER WAVE ANALOGY OF LIGHT

Arguing whether light was a particle or a water wave was getting old by the beginning of the nineteenth century when Young devised his two-slit experiment to show that light behaved just like water waves behaved.

Young took a single source of light, passed it through two slits in a barrier, allowed the light to intermingle after it passed through the two slits and collected the resulting light on a screen. The screen showed an interference pattern, a pattern of dark lines interrupting the light.

Young interpreted this to be analogous to the result of two sets of water waves intersecting each other. When water waves intersect, the troughs and peaks reinforce one another, but when the troughs and peaks intersect, they cancel one another out so that the surface of the water appears as if there had been no wave.

The dark bands of no light appeared as if there were no light, so that the peaks in the waves of light from one of the slits had to be canceling out the troughs in the waves of the light from the other slit.

"Appears" is the operative word, a word that, when applied to the absence of light produced by Young's two-slit experiment, illustrates the empirical process at its most delusional.

Water waves are a disturbance of a medium, and for that medium to be disturbed, it has to be contained by a force, gravity. Water waves are actually an interplay in the water medium of two forces, the temporary disturbance of the current force of gravity and then the current force of gravity regaining ascendancy over the temporary force of the disturbance until gravity once again constrains the contained water.

Young's experiment dealt with light which had no medium, and if it did, the medium was constrained by no force, and if it was, the force constraining the medium had not been disturbed.

More delusional, the dark lines of no light were analogized to the appearance of no waves when the crests and troughs of the intersecting water waves canceled each other out.

But when the intersecting water waves produced no waves, there was still water!

When there was no light on Young's collection screen, there was just that, no light!

The analogy of light to water waves made by Young's two-slit experiment is not only baseless, it is incomprehensible.

But the desire of Empirical Science to avoid dealing with reality for fear of being held accountable for any actual explanations of that reality prevailed.

It concluded that Young's two-slit experiment demonstrated conclusively that light was just like a water wave.

This left a nagging question about the analogy because, while light didn't exist if it were a wave, it still existed.

IF LIGHT DOESN'T EXIST, THEN THE MEDIUM THROUGH WHICH IT MOVES MUST

Water waves don't exist because they are merely the effect of a disturbance in a medium. If light was analogous to water waves, which Young's two-slit experiment proved beyond a shadow of a doubt, after all, all you had to do was perform the experiment and the interference patterns invariably appeared showing light to be like water waves, if light was a disturbance of a medium like water waves was a disturbance of a medium, then light had to be moving through a medium.

Here, Empirical Science came into its own.

If something has to exist then Empirical Science can do its job and name it, and that name for all time can be used to understand what is going on with the name becoming a substitute for reality.

For light to move from one place to another as a wave, it had to be moving through a medium which, after reference to duly constituted authority, was determined to be the aether, the ether a latter-day Sparks sent his radio waves through early in the twentieth century.

Here was something for Empirical Science to spend its time measuring. It didn't matter that it was just made-up stuff created to justify the idea that light was like a water wave. Even though light permeated the environment, was, in fact, a requirement before that environment could even be seen, was what we used to see physical reality, we didn't know what it was, so we said, it didn't exist, but because it didn't exist, the aether must, and therefore it's time to get down to the business of measuring the properties of this reality in reality, this palatable substance with a name that makes it understandable.

Measuring aether occupied wave theorists for the better part of the nineteenth century. Its tensile strength, its density, its resiliency, its pervasiveness, its constitution, even its speed and direction of motion were measured, debated and remeasured.

Its existence was used as the basis of one of the most famous experiments in history where Michelson and Morley used interference patterns produced by light being sent through the aether at ninety degree angles in an attempt to measure the speed and direction of the Earth in space. Instead of demonstrating the speed and direction of the Earth in space, the experiment demonstrated conclusively that space and time, the space between matter and the interval that occurs when that space changes, not only existed as objective quantities in reality, but were variables, a fitting result of creating the fantasy of aether in the first place.

The experiment really demonstrated that the aether didn't exist!

A SEPARATE COURSE OF REALITY

While wave theorists were noodling aether and creating physical time and space, some people were fiddling around with physical reality, using the process of trial and error to uncover actual facts.

People had been parlor playing with magnets and electrostatic gimmickry for centuries. However, it was not until Volta produced a steady flow of electricity with a battery around the same time Young was jerryrigging his two-slit experiment to prove light was a wave that the play could become serious.

PLAYING WITH ELECTRICITY AND MAGNETS

Once a steady flow of electricity was achieved early in the nineteenth century, the first thing that was noticed about electricity and magnets was that the flow of electricity in a conductor would deflect a magnetic compass needle.

We had (and have) no idea whatsoever what was flowing through the wire, and we had (and have) no idea whatsoever why a compass needle lined up with the Earth's magnetic field (for which we had and have no explanation), but we could see that putting a compass needle next to a conductor with a flow of electricity in it deflected the compass needle.

This led to several decades of playing around with conductors and magnets, and resulted in the observations that resulted directly in our electric generation based civilization.

One observation dealt with the movement of a conductor and a magnet. Magnets have lines of force which are easily measurable by dusting a piece of paper with iron filings and letting the magnetic lines of force pass through the paper to organize the iron filings. If the conductor was closed in a circuit, then moving it at right angles to the measurable lines of force that surrounded a magnet resulted in an electric current.

Cutting magnetic lines of force at a right angle with a closed conductor somehow produced a flow of electricity that was the same electricity Volta's battery produced.

What would happen if an electric current carrying conductor rather than a magnet were used to influence the iron filings? A flow of electricity passing through a conductor placed perpendicular to the paper containing the iron filings caused the iron filings to line up in concentric circles around the current carrying wire.

This meant that a flow of electricity had lines of force surrounding it similar to a magnet.

What would happen if a conductor in a closed circuit were to be moved at a right angle through the lines of force produced by another conductor carrying electricity?

Joseph Henry demonstrated that passing a conductor in a closed circuit at a right angle through the lines of force produced by another conductor carrying electricity produced an electrical flow in the conductor, releasing the genie that produced the miracle of modern civilization.

WHAT'S GOING ON WITH MAGNETS AND CONDUCTORS?

We don't know what electricity is and we don't know what magnets are but we do know that we can produce electricity by moving a conductor at a right angle through the lines of force produced by a magnet, or in the alternative, at a right angle through the lines of force produced by a conductor.

What we got was a total violation of the law dealing with the conservation of energy.

We got energy appearing out of nowhere!

This little problem was easily overcome by the empirical process of naming, calling electricity *a form of energy* and equating the amount of work electricity could produce when electric motors were fabricated to the amount of work a horse could do. If we have a one horsepower motor, we know what we are dealing with. We don't have to worry about where the energy is coming from. Electricity is one of the forces of nature, and we have just learned how to tap into that force.

Who needs to understand so long as we can know simply by naming things?

WHAT DO WE KNOW?

The lines of force that surround a conductor that contains an electric flow are called inductance.

As opposed to gravity and light, which are obvious, inductance is the most important nonobvious phenomena in our existence.

Instead of holding inductance up for examination, asking what it was, where it came from, how it worked, what caused it to produce electrical flows, Empirical Science analogized it to the lines of force around a magnet.

There is nothing wrong with this analogy, for the two are indeed the same thing.

However, by analogizing inductance to magnetic lines of force, Empirical Science answered all of the questions it had about inductance because it had already answered all the questions it needed to ask about magnetic lines of force.

Magnetic lines of force came with magnets. Magnetic material produced a disturbance in the great whatever that evoked these lines of force. All that was happening in a conductor was a different way to evoke lines of force out of the great whatever.

What else did Empirical Science need to know?

If instead of analogizing inductance to magnetic lines of force Empirical Science had analogized magnetic lines of force to inductance, then it would have left itself open to the question of where these lines of force came from, something that would then take mental effort to solve.

Mental effort is not empirical because the mind doesn't exist, and if it did, it couldn't be trusted. Better to leave it up to predictive facts to prove ideas.

Better to use the mental effort to derive the properties of that which doesn't exist, the aether, angular momentum, straight-line motion, than to attempt to solve problems of physical reality, things like what makes dropped objects move, planets rotate and orbit, and what shines from the sun.

A SEDUCTIVE CONCEPT

One question could have clarified the nature of inductance.

As electrical flows were started and stopped, their associated inductances were established and collapsed around the flows. If moving a conductor through an inductance could cause a flow of electricity, could a flow of electricity be produced by allowing establishing and collapsing inductances to pass through a conductor?

Establishing and collapsing inductances did, indeed, produce a flow of electricity in conductors maintained at right angles to the conductor establishing and collapsing the induction.

This little trick couldn't be duplicated with a magnet because magnetic lines of force were permanent. It is the starting and stopping of the electrical flows that establishes and collapses the inductances.

There was no way to establish or collapse the magnetic lines of force around a magnet.

Magnetic lines of force simply existed!

This physical limitation should have eliminated magnetic lines of force from the playing field, reducing them to a historical curiosity along the lines of electrostatic generators.

But the idea of establishing and collapsing inductances implies a motivating force at work, something that can actually accomplish something, move something in physical reality, evoking tantalizing fantasies of some self-perpetuating cosmic water pump.

We have no idea at all where the electricity is coming from, but it is capable of doing work. It is very seductive to think of work being performed by collapsing and establishing inductances. We don't know what inductances are, we have already confused them with magnetic lines of force whose explanation also escapes us, but the whole business results in work, so the idea of the movement that occurs when inductances are established and collapsed is something, something happening in reality, something we can wrap our minds around.

So Empirical Science performed a little bit of reductionism, a practice dear to the hearts of its faithful.

An electric field, the flow of electricity, produces inductance, a magnetic field.

A magnetic field, the magnetic lines of force, produces an electric field, a flow of electricity.

Thus, every field, magnetic or electric, induces a corresponding electric or magnetic field, giving rise to Maxwell's equations describing electromagnetic waves.

The mind just rolls along in the universe as it follows wave after wave, pushed first by a collapsing magnetic field establishing an electric field, and then pushed by the collapsing electric field establishing a magnetic field which in turn collapses establishing an electric field, on and on forever.

The sophistry of this twaddle is self-evident.

First, defining everything as a field is absurd. A flow of electricity, the inductance produced by the flow and magnetic lines of force are at least two different things and the last two are also distinctly different in their operation.

Second, while inductances are like magnetic lines of force, magnetic lines of force are not like inductances in that they do not collapse and establish themselves. They simply exist in proportion to the magnetic material with which they are associated.

Third, while an electric flow actually induces an inductance into existence whatever that inductance might be, to get electricity out of magnetic lines of force something has to move the conductor back and forth at right angles through the magnetic lines of force. It takes work to get electricity out of magnetic lines of force which means it isn't induced.

Thus, half of the statement described by Maxwell's equations is wrong on its face. A magnetic field does not collapse and therefore does not itself induce an electric field.

Without an electric field induced by a collapsing magnetic field, Maxwell's equations produce the physical specter of one hand clapping, and the fantasy of a motive power pushing electromagnetic waves through endless space collapses.

NOTHING BECOMES A SMALL PART OF NOTHING

Maxwell's equations contemplate electromagnetic waves created as a result of the interaction between collapsing electric and magnetic fields, with wave lengths dependent on the rate at which the fields interact. A slow drop in a pool creates leisurely waves, while a fast drop creates faster waves. A drop's effect on the pool is soon dissipated as gravity overcomes the force of disturbance. With electromagnetic disturbances, however, the motive power is the interaction of the collapsing fields, which produces a wave which continues forever.

Thus, once generated, electromagnetic waves are self-perpetuating, with the alternating collapsing electric and magnetic fields pushing the wave through the aether until it splashes onto the end of space at the beginning of time.

In seeking to determine the properties of these waves, Maxwell, computing the speed they would have to be moving as a result of their alternating collapsing fields, predicted that they would move at the same speed light waves moved. Because the speed of light had been measured some two hundred years earlier, the prediction was akin to Galileo's mathematical prediction of the optimal angle for distance of cannonball fire proved by Tartaglia half a century earlier.

Not much chance of error!

The success of the prediction allowed Maxwell to view light as an electromagnetic phenomenon.

In fact, light was just an insignificant portion of the electromagnetic spectrum, one we gave entirely too much importance to simply because light is how we subjectively see the world.

Thus, the decision that light was nothing was a correct decision, because even though it was something, it was only a small part of that something which, being a wave, was, in turn, also nothing, merely a disturbance of a medium.

Maxwell's vision of a self-perpetuating oscillating electromagnetic field moving forever through space, while seductive, falls short of reality in that it requires a motive power that is unnecessary, an interaction of two aspects of the same thing that cannot occur because magnetic fields do not collapse and still needs a medium which doesn't exist.

HERTZ DETECTS ELECTRIC OSCILLATIONS AT A DISTANCE

There is, however, a more elusive prediction on which Empirical Science bases its belief that light is an electromagnetic wave.

The fact that a Leyden jar, an early electrical storage device, could be used to produce electrical oscillations led Hertz to wonder if he couldn't devise something to detect the frequencies produced by those oscillations some distance away from the Leyden jar.

Although Maxwell's equations had been promulgated at the time of Hertz's trail and error effort to detect electric oscillations over a distance, Hertz did not believe that detecting the resulting frequencies over distance would prove Maxwell's equations. Rather, he stated that "I am never content until I have constructed a *mechanical* model of the object that I am studying. If I succeed in making one, I understand; otherwise I do not. Hence I cannot grasp the electromagnetic theory of light. I wish to understand light as fully as possible, without introducing things that I understand still less. Therefore, I hold fast to simple dynamics for there, but not in the electromagnetic theory, can I find a model."

Reasoning backward from the fact that a collapsing inductive field produced a flow of electricity, he devised a detector that would measure an induced flow in a circuit some distance from the source of the oscillations producing the frequencies.

Succeeding, Hertz went on to show that these frequencies behaved just like light, they reflected, refracted, even diffracted and moved at the same speed.

EMPIRICAL SCIENCE DRAFTS HERTZ'S DETECTION AS PROOF OF MAXWELL'S EQUATIONS

The logical conclusion would have been that because frequencies produced by electrical oscillations were similar to light, that light must therefore be a frequency,

However, as reluctant as Hertz was in referring to light as electromagnetic waves, leading to his relative obscurity in relation to his incredible contributions, his heirs, with Maxwell's seductive fantasy in mind, concluded that because everyone knew that light traveled in waves, the frequencies produced by a Leyden jar or an oscillating inductive coil were also waves.

As they were waves, then Maxwell's equations were proven because they predicted Hertz's discovery, and as Maxwell's equations were proven, then light, in Maxwell's words, "consists in the transverse undulations of the same medium which is the cause of electric and magnetic phenomena."

And, of course, because Maxwell's equations predicted Hertz's discovery, Empirical Science claims sole credit for the television industry.

AN ASSUMPTION THAT TRAVELS FOREVER

Just because light is but a small part of the electromagnetic spectrum doesn't mean that it is immune from the immutable laws applicable to the electromagnetic spectrum. If it is a part of the electromagnetic spectrum, then like the electromagnetic spectrum, light can travel forever under its own motive power of alternating collapsing magnetic and electric fields.

But light doesn't travel forever!

While there is no question that frequencies can be produced by the interaction of inductive fields that propagate an expanding sphere away from the oscillator producing them, whether it be a Leyden jar or an inductive coil, light diminishes inversely with the square of the distance over which it travels.

In fact, the oscillations that have been used to carry radio frequencies also diminish with the square of the distance over which they travel.

And yet, no matter how many textbooks on astronomy, optics or physics are produced, many with identical illustrations and cloned, precisely misworded phraseology, they only rarely contain a discussion of the fact that light diminishes with distance, and then as a statement rather than as a discussion. Diminishing light is treated as a curiosity in the fashion of the H. G. Wells measurement of gravity, where if mass/gravity were correct, then the center of the Earth would be mathematically weightless.

When your laws, principles and scientific facts have no basis in reality, the opposition of reality to those laws, principles and scientific facts become mere curiosities.

But diminish light does, so, as Empirical Science views the light from stars at the end of the universe and the beginning of time, it somehow has to take into consideration the fact that light diminishes, if only to give the fact lip service.

Because Empirical Science believes that light can travel forever as a part of the electromagnetic spectrum, it has to carve out that part of light that is obviously diminishing.

Since ancient times, stargazers have graded stars by magnitude, with first magnitude stars being the brightest. For thousands of years, the assumption was that the brightest stars were the closest, and the less bright stars were further away until there were stars that could hardly be seen at all, which meant their light had expanded almost out of existence. This assumed stars were all the same size and burned at the same rate.

However, as telescopes were developed and improved, dimmer stars were brought into focus better, a process, one would think, that was the natural result of the telescope lens concentrating less light better.

But, instead of taking this straightforward point of view, the delusion crept into the empirical process that the stronger the telescope, the closer it was moving us to the stars. Thus, instead of a telescope concentrating light, it was seeing further into space. When Maxwell's equations grafted light onto a wave that traveled forever, the size of stars became problematical. It wasn't long before a near bright star was resolved by a powerful telescope into a galaxy of stars located very far away. If one star didn't obey the dimmer the star the further the distance dictate, then it was not a dictate, it was a canard like the canard about the sun going around the Earth, and had to be totally discarded.

Brightness was therefore something that was only apparent, like light. It didn't itself provide us information about comparative distance because it didn't take into consideration size and luminosity, the brightness resulting from the star's energy output.

If the stars weren't all the same size, with distance producing dimness, then the stars were all different sizes and burning with all sorts of different intensities.

The fact that Empirical Science was dealing with stars of different size, energy and intensity could be empirically proven right here on Earth.

A hot ember will look the same as a cooler ember if it is far enough away.

A large cool ember will look the same as a small hot ember if it is far enough away.

And, of course, a cool ember can be measured to be older than a hot ember.

As no one can disprove the claim that stars are different sizes or burn with different intensities, it is a safe assertion that all stars are different sizes and burn with different intensities.

It is a star's magnitude, its brightness, its luminosity that is diminishing with distance. The brightness itself is an illusion resulting from size, distance and inherent luminosity. But because the brightness diminishes inversely with the square of the star's distance, we can use this mathematically measurable fact to compute how large a star is, how intense that star is burning and by golly, even how old the star is.

Light might be part of the electromagnetic spectrum which travels forever, but brightness is intrinsic to a star, and it is this brightness that diminishes inversely with the square of its distance that allows us to further categorize the star as to size, energy and age.

The measurable fact that light diminishes doesn't conflict with the belief that light is a part of the electromagnetic spectrum and electromagnetic waves travel forever because it is the brightness, the luminosity, the magnitude that is diminishing!

This absurdity, where a measurable physical phenomena is ignored so that Empirical Science can see what doesn't exist, light at the end of the universe and the beginning of time, but employed to allow Empirical Science to measure that which can't be measured, the size, energy and age of stars, clusters and whole galaxies, heck, even the end of the universe and the beginning of time illustrates the delusional nature of believing that finding a fact predicted by a belief proves the belief.

The basic misconception, that light can travel forever while actually diminishing, illustrates an even greater misconception, the drafting of facts that appear similar to beliefs into the service of those beliefs as predictive facts, whether it be Hertz's detection of electrical oscillations at a distance to prove light is an electromagnetic wave or Young's jerryrigged two-slit experiment to prove that light is a wave to start with.

But then, when we are dealing with beliefs, made up stuff, anything goes!

LIGHT *IS* SOMETHING

The consequences of not knowing what light is are conceptually catastrophic. We not only see physical reality with light, Empirical Science uses light to measure, or at least, to think it measures, all sorts of things, starting with the distance to other stars, continuing with what stars are made up of and concluding with which way the sun is moving in space, measurements which are all erroneous, with the latter error being mathematically demonstrable. (See the cover of At the Gates of the Citadel, the first volume of The Copernican Series.)

The belief that light does not exist, is not, in fact, something, but rather a disturbance of a medium, was not even shattered by another of Hertz's discoveries. Hertz's observation that light not only behaves as a particle, but is interchangeable with electricity, the photoelectric effect, should have put the wave silliness to rest once and for all.

But how can the wave silliness be put to rest in a system where beliefs are easily converted to facts by simple belief. Pass light through two slits and see the interference patterns. Light is a wave.

Bell rings, lick lips.

When an experiment is devised to demonstrate a fact, as Young's two-slit experiment was, then it becomes extremely hard to sit down and figure out just what is happening when light is divided, recombined and collected on a screen, especially when we don't think of light as something that diminishes uniformly with distance until it diminishes out of existence, but rather think of it as nothing at all.

Knowing that light is a wave from viewing Young's two-slit experiment, and knowing that light is a particle from the demonstrable physical output of turning light into electricity, Maxwell's heirs have concluded that light is both a wave and a particle. What more fitting conclusion to the conception of light as some sort of incomprehensible jumble of waves moving at right angles to each other throughout all three hundred and sixty compass degrees, and all points in between, than its conception as a nonexistent wave that is also an inconceivable particle, a particle having both location in space and movement in time, but never simultaneously.

The mistake, of course, is the hallmark of Empirical Science, assuming away that which it is too lazy to explain.

When it couldn't come up with a force causing the Earth to rotate on its axis, Empirical Science assumed the force away, making the force a historical force which it named inertia. When the motion of the planets couldn't be accounted for, Empirical Science assumed the force causing that motion away as straight line motion occurring as a result of a force exerted at some time in the past. When the mechanics of gravity could not be described, Empirical Science assumed away the force that makes all dropped objects move by claiming it was a property of the matter toward which the objects were moving.

When it came time to explain what light was, Empirical Science assumed its very existence away, calling it nothing more than a ripple in the occult pond of the universe.

And when Empirical Science assumed light away by analogizing it to a water wave, it assumed away any coherent explanation how light is produced, what causes it to exit physical matter at one hundred eighty-six thousand miles per second, how it can move through space in a way that it not only diminishes with distance, but does so precisely, inversely with the square of that distance, and more incredibly, does so uniformly.

When Empirical Science assumed light away, it assumed away any explanation for how light interacts with matter, how it not only bounces off of matter but interacts with the surface of that matter to produce heat, and how, if there is enough light, it causes that matter to itself start emitting light.

And, because in assuming away light, Empirical Science has eliminated coherent explanations for all of these things, it has also assumed away any reasonable explanation of what matter itself is, how it is made up, how it got that way, and how it can, itself, move with no apparent motive power as a falling object or as a rotating or orbiting planet or as light itself.

The ignorance that empirical laziness produces is insatiable. It consumes all available comprehension.

It's time to start asking the hard questions, and take the time to answer them, that will allow us to start to understand what the hell is going on around us, why objects, from rocks to planets, move, what the universe is and what we are.

It is only then that we might start to get a glimmer of what our place in the universe is.

Instead of assuming light is nothing, we have to assume that light is something, and then get busy finding out what that something is.

It's simply silly for Empirical Science to pretend that it knows what's going on at the end of the universe and the beginning of time when it doesn't have the foggiest idea what's going on right under its nose.

Unless, of course, it's practicing empirical misdirection, hocus pocusing the nondisprovable to call attention away from its abysmal ignorance of the most obvious phenomena in our existence, making Empirical Science a flimflamming freeloader of miraculous magnitude and catastrophic consequences, because, thinking we know everything, we will become extinct knowing nothing.

7. FOUR HUNDRED YEARS OF PROGRESS: FROM BEING AT THE CENTER OF THE UNIVERSE TO BEING ISOLATED AND ALONE IN THE UNIVERSE

"In the whole of sidereal astronomy there is, perhaps, nothing more difficult to determine than the parallax of a star. To think that among all the stars in the sky there is not one which shows a parallax of *one second*! Now, one second is a millimeter seen at two hundred meters, it is a hair of a twentieth of a millimeter seen at 10 metes (32.8 feet)! Well, it is in this width that the annual motion of a star is performed. The telescope magnifies it, of course; without this it would be absolutely imperceptible; but how easily it can be concealed by the imperceptible motions of the telescope, by the influences of temperature, by refraction, precession, nutation, aberration, and by the proper motion of the star itself in space! All these united influences amount to *several seconds*, and are themselves subject to some uncertainties, and instrumental errors must still be added to them. How, then, shall we extricate trustworthy indications of the minute displacement due to the effect of the earth's motion? *Astronomers, however, succeeded in doing so for some stars.*" Flammarion, *Popular Astronomy* (1894)

SMUG SCIENCE

Nothing is more humorous than listening to some modern empirical slug snicker with condescension at the thought that humans could be so arrogant as to believe the ancient notion that they were alone, occupying the center of the universe. Broach the subject of UFOs, however, and the condescension will turn to outrage as the slug summons the empirical police.

Bruno was probably burned at the stake more for his suspected double agent activities than for his scientific beliefs, but there is no question that the logical conclusion he drew from what he accepted as the Copernican reality of the Earth orbiting the sun was disquieting to his interrogators.

If, Bruno reasoned, the Earth orbited the sun, then the lights we saw dotting the skies were also suns and they also had Earths orbiting them and those Earths also had people on them just like us.

The primary reason that this viewpoint was disquieting four hundred years ago was the implication that if there were other people on other planets, then those people's existence wasn't covered by the creation story in the bible. Such heretical assertions provided an excellent basis for frying an intelligence agent who may just have gone off-center.

Religion aside, there can be no greater reality than the truth of Bruno's assertion. The stars are suns, and those suns certainly have planets and those planets certainly have people on them.

So, when millions of people sight UFOs in the sky that, to empirical strictures, are aerodynamically impossible, but which, in this day of camcorders, are recorded traveling over courses extending for hundreds, sometimes thousands of miles, why not just accept them for what they are, vehicles occupied by other people representing advanced civilizations from neighboring star systems who are dropping by for a look-see at the local backward natives?

Why would Empirical Science insist on following a four hundred year old literal interpretation of the Bible?

It's certainly not because Empirical Science doesn't believe that life on other worlds is possible. NASA recently rigged up some life on Mars to give manned interplanetary exploration a long overdue shot in the arm, and astronomers, having exhausted other idiocies to garner publicity, are running up each other's rear ends in an attempt to show that mass wobble of stars is not only detectable, it demonstrates that planets orbit neighboring stars.

Is it because Empirical Science demands that its hypotheses be testable, and that life in other solar systems is merely another hypothesis that requires testing? Is it a perversion of the putative scientific method, a claim that until some evidence, more than a shred of evidence at least, turns up to demonstrate intelligent life on other planets, that the hypotheses of other intelligent life in the universe remains untested?

The incessant sightings of UFOs put Empirical Science at risk because it either has to claim that the population is suffering from mass hysteria or it has to deal with the possibility that the sightings are substantial, and that its laws therefore do not in fact describe physical reality, that Empirical Science is, in fact, what it is, a misassembled pile of occult crap.

Empirical Science can't prudently make either claim, that we're dumb or that it's dumb, although recently it has been coming perilously close to claiming we're dumb.

So Empirical Science makes the claim that UFOs can't exist because their behavior doesn't conform to the laws of nature as those laws are presently understood.

This is so patently absurd that it is laughable. Even Empirical Science is embarrassed by its claim that because it doesn't understand what's going on, what's going on isn't going on.

Therefore, Empirical Science has to fall back on its final line of defense against the ultimate apostasy, that we are being visited by curious neighbors in the universe who know something Empirical Science doesn't know, by claiming that the distance between the stars is so great that even if advanced civilizations exist, interstellar travel would be impossible.

In short, we may not be the center of the Universe but we might as well be, because there t'aint nothing out there that can drop by and say hello (as opposed to nothing out there that would want to drop by and say hello).

Now this is an extraordinary claim, that we can look out at the pinpricks of light in the darkness of space and measure the distances to those pinpricks, and in doing so determine they are so far away that traveling between them is impossible.

One must always be wary of extraordinary claims.

In fact, extraordinary claims require extraordinary proof!

We wouldn't want to accept just any old proof or even tenuous proof that the distances to the stars are measurable and so vast as to be impassable because the psychological consequences of such a belief are just too great, dictating as it would a determination whether we are actually sophisticated beings curious about the universe, or just savages isolated by ignorance and superstition on some island in space, a dot of green and blue whose inhabitants' development, stunted by some unfathomable ancient cataclysm beyond their ken, has left them a sociological curiosity to curious voyagers.

WHY WOULD WE TRICK OURSELVES INTO BELIEVING WE KNOW SOMETHING WE CAN'T KNOW?

Questions of perspective plague the mind, which attempts to fill in the missing elements of the pictures of reality it forms by what it has available in its recall. If we are looking out to sea and we see a ship all alone bobbing up and down on the waves, we have no way of knowing how big the ship is. If our job is to repel invaders, knowing the size of the ship would be extremely important.

However, if the atmosphere is extremely clear, what's far away might appear to be up close, and if the day is hazy, what's up close might appear to be far away. We evolved with two eyes so that we might have depth perception as we moved through reality, but that depth perception disappears at distances where movement is not contemplated and thus strategies to deal with attempting to determine size over distance couldn't evolve.

We are really in the dark about how big the ship is because we have no way of determining how far away it is.

We are driven to estimate the size of the ships, however. Knowing how big the ship is would tell us how many people it carried, and thus the size of the invasion force.

Our survival forces us to make the estimate.

If we say the ship is a long distance away, we might say that the ship is very large to look as large as it does. We do some calculations about how big "large" could be and decide that the ship must be ten miles away because the largest ship that could be floating out there would be a thousand feet long, and a thousand foot long ship would be the size we measure the ship to be if the ship were ten miles away.

That circular computation completed, we can say that a thousand foot ship would hold a thousand people.

Of course, if it was clear, the ship might only be a mile away, making it only a hundred feet long, holding a hundred people.

Or, if it was hazy, then the ship might be a mile away and still be a thousand feet long, holding a thousand people.

All of this stuff is just mathematical bullshit because we simply don't have enough information to produce measurements. Businesses play what-if all the time with software programs that manipulate variables. The Office of Management and Budget will provide what-if disks dealing with the national budget so that people can make their own assumptions.

All agree that in things such as the budget, or business decisions, variables represent a future no one can know.

On the other hand, computing the number of people on the ship is not a crystal ball operation. There are a number of people on the ship, the ship is real, it exists, it's out there, it's a current fact, it's a threat to our continued existence and we have to come up with a number.

If we are like General McClellan, the eminently unsuccessful civil war general who consistently insisted on overestimating the enemy presence, we can overestimate the number of people on the ship and go into vigorous retreat. Or we could underestimate the number of people and meet the enemy totally outgunned.

When we are worried about invaders, we are forced to make assumptions about reality and then base estimates of the quantitative nature of that reality on those assumptions.

But why make an estimate if we don't need to know?

The distances to the stars and galaxies and the number of stars in a galaxy are not things that provide us with a present threat.

Why would we want to compute the distance to the stars or the number of stars in a galaxy if we didn't have to?

Why should we produce mathematical constructs to fool ourselves into knowing something that we can't know?

DISTANCE BEGETS STARS

So that we will know what we are up against when we explore how Empirical Science has come up with its conclusion that even if there were other civilizations in the universe, we are alone because the distances are so great between the stars that they could never be traversed, it is necessary to look at the mechanisms of astronomical calculation.

Just as the number of troops on the bobbing ship depends on the size of the ship, so does the number of stars in a galaxy depend on the size of the galaxy. Astronomers are used to saying all sorts of silly stuff because they know that no one can double check them. Thus, an astronomer can point to a tiny dot in the sky and say that's a galaxy, an island universe, and it's made up of two and a half billion stars.

We, simple fools that we are, gasp at the massive brain work that must have allowed such a godly creature to be able to asseverate such a fact. Two and a half billion stars! That's a whole lot of stars.

We wouldn't think of asking this heavenly purveyor of fact how he had the time to count all those stars. We know, because we are told over and over, that Empirical Science is based on empirical observation, so we know that someone, somewhere, must have sat them self down and done the counting, and a dozen more recounted to empirically ensure that the original count was correct, but we don't dwell on how this might have been accomplished.

And, of course, we don't even dare to think the astronomer is just talking through his hat, making up bullcrap answers, that he hasn't the foggiest idea how many stars there are in the galaxy, in fact, a hundred years ago didn't even know there was a galaxy in which to count the stars.

What the astronomer has done is the following:

He has estimated the distance the galaxy is from Earth.

Using the estimated distance, he has estimated the galaxy's size.

He has then estimated how far he thinks an average star in a galaxy is from another average star.

He has then determined the volume of the galaxy using its estimated size.

He has then divided the average volume a star occupies if it is an average distance from all other stars into the average volume of the galaxy as determined by its estimated size as determined by its estimated distance, and come up with an estimated number for the stars.

Of course, this hasn't all been a bed of roses. The astronomer has no way of knowing how many stars make up the center of the galaxy, he just knows that the number of stars per volume of space in the center of the galaxy is greater than the number of stars for the same volume of space in the arms because he can't resolve the individual stars in the center of the galaxy. So he has to create one estimate for volume at the center of the galaxy and another for the arms of the galaxy.

But that's what Empirical Science is, a process of making exacting computations based on absolutely no fact whatsoever. One actual mathematical computation, volume into volume, gives the whole fantasy the color of measurable reality and thus deludes Empirical Science into believing it is dealing with measurable fact.

So, with all of these estimates based on distance, what does distance do to the estimates?

If the astronomer underestimates the distance to the galaxy, the galaxy is further away than it looks, it is larger than it looks, and therefore underestimating distance reduces the number of stars in the galaxy.

Simply by adding a little distance to his estimate of how far away the galaxy is, the astronomer does what all good necromancers do, he creates matter out of nothing.

On the other hand, in the unlikely event that he overestimates the distance to the galaxy, then he is creating too many stars in the galaxy. Reducing the distance would take stars away from his estimate, destroying matter as it were.

Why does the astronomer do all of this, make up answers for things about which he has no knowledge and for which he can have no knowledge?

What would you do if you were asked a question the answer to which you had not the foggiest idea and for which there was no obtainable answer to start with?

Would you want to say, "Hey, I don't know the answer to that question. It's just one of those things that we can never know. We are in a galaxy, and we have no way of knowing how many stars make it up. Other galaxies are blurs of light. We can't tell how far away they are, we can't tell how big they are and we can't tell how many stars make them up."

"Well, boot me Mama," we reply. "Didn't this idiot want a new telescope? What good's a new telescope if he can't answer our questions?"

The astronomer is in danger, at a minimum, of losing his toys if he comes back with the truth, that he just doesn't know the answer to our question.

So he's going to come up with an answer. Stupid us don't care what the answer is, we just want to hear an answer. Two and a half billion stars is about as meaningful to us as three and a half billion stars.

But why not a more realistic estimate of the number of stars?

If you were going to hand out stupid answers, would you hand out something that was questionable?

What if a sensible answer were in the tens of thousands?

That's something the average idiot can understand. We all spent time trying to count to ten thousand when we were kids and learning how to count. We might go out and second guess the astronomer if he came up with a reasonable answer to how many stars there were in a galaxy.

Why should he bother coming up with a reasonable answer when there is no answer to start with? Why not just overestimate what can't be estimated, and come up with mind-boggling billions instead of comprehensible thousands?

CREATING VAST DISTANCES HAS VAST PSYCHOLOGICAL CONSEQUENCES

Before describing the step by step descent that led to the absurdity that distances in space are so vast that they can never be traversed, the consequences of such a belief deserve comment. When we attempt to make choices as to what we are going to believe, when we contemplate whether we can perhaps do something that will carry us beyond ourselves and our present capabilities, the choices, in many cases, have already been dictated by a priori blindness.

It's fairly obvious that if we think that distances in space are so vast that they can't be traversed, then we will never attempt to traverse them. But what does that mean in immediate terms?

The first thing that it means is that this is all there is because there ain't no more. We live on the third planet from the sun and none of the other planets are even remotely habitable by our requirements. If something happened to the Earth, then that would be it for us.

Because our life is dependent on our environment, without that environment, with an Earth like Mars or the moon, we would be no more.

If our existence is dependent on the Earth for its continuance, then we wouldn't willingly want to deal with the end of the Earth's existence. If we cannot bring ourselves to deal coherently with our own end, we can't be expected to deal with the end of life on Earth. This would drive us to create an Earth that would be habitable forever in the future, or at least for so long that its end would provide us with no immediate misgivings about our own end.

Thus, dipping into the old billions pool where the concept of the number is impossible to comprehend, we could say that the Earth has been around for billions of years, and its going to be around for billions of years to come.

When we have ourselves isolated from the rest of the universe, and blessed with life on a planet that will continue to exist like it is for billions of years into the future, an empirical substitute for the nonempirical belief in eternal life, we can settle into the same mindset we had when we were at the center of the universe.

What mindset is that?

Pure, unadulterated arrogance.

Once again we are the chosen people, the life form God created to populate the universe, the universe, of course, being our universe. Certainly there might be other life forms out there, and even the smug Carl Sagan, for whom all evidence that contradicted belief was no evidence at all, dearly hoped to encounter evidence of it someday, but for now, it's just out there somewhere and of no concern to us.

Arrogance dealing with our place in the universe leaves us without aspirations. We focus our attention on the technology required to feather our nest rather than on the technology required to extend our range of survivability into the future.

Our minds evolved to move us safely through the environment by forming a picture of that environment so that it could be compared with the picture of the environment we had stored in our recall. A by-product of this evolved mechanism is ..the ability to form a picture of the environment when it isn't there, and because we can alter the pictures we form of reality when reality is not present, we can alter reality. This accidental result of evolution allows us to create pictures of reality that are better suited to our survivability than actual reality, and to then go out into reality and alter that reality so as to make physical reality more conducive to our continued existence.

While the mind evolved to move us through reality, it is the tool we have to extend our range of survivability in that reality by forming pictures of that reality, altering those pictures, and then altering physical reality to conform to those pictures.

If we are nested, if we are happy, if we are content, if there is no threat to our continued existence, we will not attempt to alter the pictures we have of reality.

In essence, believing that we are isolated from the universe, that we cannot leave this solar system, that we are stuck right here on this planet forever causes us to place all of our eggs, our genes really, in one basket. We have no idea how long that basket, the Earth, has been here, we have no idea how long the solar system has been here and we don't know how long the Earth is going to accommodate our existence.

But simply by believing that we are unique, isolated, alone, we have accomplished our own demise, we have acquiesced in the probability that the unique life form that evolved on this planet, orbiting this star, in this galaxy, in this small corner of the universe will die in place where it evolved, never having made a rational attempt to determine what its place in the universe is, or indeed, whether it even has a place in the universe.

ROBERT HOOKE BECOMES THE FIRST ESTABLISHMENT HACK

In mid-sixteenth century England, with the ascension of astronomy buff Charles II to the throne, Oxford scholars felt that it was necessary to rescue British Science from Cromwell's revolutionary disorder in the face of intellectual onslaughts from competing nations such as Italy, France and the ruling power of Holland. Science had become a big gun in the war between nationstates as the competition for colonization of the planet continued apace. Reputation was the report of that gun, with the larger the reputation, the greater the report and the bigger the implied gun.

Scientific firsts, especially discoveries and proofs of hypotheses, were the handmaiden of reputation, bringing it forth in all its pristine glory.

Mistakes about scientific firsts, on the other hand, were the bane of reputation, with false claims as to discoveries and proofs placing a nationstate in danger of becoming a laughing stock in the face of the world.

The Oxford scholars sought to form themselves into a Royal Society to organize Britain's discoveries and proofs, to filter them in order to avoid national embarrassment and to trumpet discoveries that were clearly verifiable, take credit for claims that could never be disproven and, if possible, claim other's discoveries as Britain's own.

Hooke, educated but poor, became the society's paid curator of experiments.

At this time, in the 1660s, just over a hundred years after the deathbed delivery of the Copernican proposition that the Earth moved around the sun rather than the converse, the question of proof was still raging. Galileo's idea of an inherent inertia turning the Earth in frictionless space had been more or less discredited in the face of his failure to realize that the tides would have long since caused the Earth to stop spinning.

There was a competition among nations to produce the Copernican proof and the nation that could prove that the Earth went around the sun would enhance its reputation beyond all bounds.

Royal Society members therefore began to kick around the idea of measuring parallax to prove that the Earth was in orbit around the sun.

WHAT IS PARALLAX?

If you hold a finger in front of your nose, you have three things to consider with respect to your finger. You have your head which contains your eyes which allows you to see your finger. You have your finger. And you have the background behind your finger. If the background is stationary, then you can move your finger back and forth side wise and see its position change with respect to the background.

However, there is another way that you can see your finger move back and forth with respect to the background that doesn't involve moving your finger. If you alternately open and close each eye, your stationary finger appears to move back and forth against a stationary background.

This movement is caused by parallax, an apparent change in position of an object because of the change in the angle at which the object is viewed. Each eye, located inches apart, views the finger at a different angle.

The members of the Royal Society discussed the fact that if the Earth went around the sun, then opposite sides of the resulting orbit would produce two angles of view. All that was needed was a sort of finger against which the two angles could be measured, and once measured, the measurement could be used as proof that the Earth did in fact have an orbit and that orbit was around the sun.

At the time, planets were determined to be planets because they moved independently against the starry background. Stars, on the other hand, were starting to be considered distant suns, with relative distance determined by brightness. Because light diminishes inversely with the square of the distance over which it travels, bright stars were thought to be up close while dim stars were thought to be far away.

The fact that some stars were close and others far away led the members of the Royal Society to wonder whether the bright close-up stars wouldn't change position against the dim far away stars when the Earth was on opposite sides of its orbit, just like a finger appeared to move against a stationary background when the eyes were alternately opened and closed

Hooke was instructed to complete measurements of the parallax of an appropriate star and report back the results.

SOME OBVIOUS PARALLAX PROBLEMS

When we run through a rain storm, the drops look like they are moving away from our direction of travel even though they are dropping vertically. Thus, viewing stars from the platform of a moving Earth is thought to cause stellar aberration, a displacement of the stars as a result of the Earth's motion. Any measurement of a star's movement as a result of the changed perspective from the Earth being on opposite sides of its orbit would have to account for this stellar aberration.

Another small problem in measuring the apparent movement of stars because of a change in the Earth's orbital position involved what any schoolchild sees when looking at stars. Stars twinkle. Because of a star's distance, its light is being refracted at continually different angles in the atmosphere so that it is first here and then there. Stars appear to move while we look at them regardless of the movement of the Earth.

With four or five other problems, the star's proper motion in space, supposedly measurable, the reliability of the telescope to name two, a correction of a second or so (one thirty-six hundredths of a degree) must be made to claim accurate measurements. Because no measurement of parallax ever approaches a second of a degree, this is quite a trick.

Hooke thus had his job cut out for him.

He was being asked to measure that which couldn't be measured!

SOME LESS OBVIOUS PROBLEMS WITH PARALLAX

The Royal Society directed Hooke to make the parallax measurements to prove that it was the Earth that was traveling around the sun and not the sun that was moving.

Well, if it wasn't the sun that was moving, then the sun wasn't moving, was it?

So right out of the chute the proceedings were handicapped with the assumption that the orbit used to define the parallax measurement was not itself moving.

The sun moves upwards of three million miles per year. In the half year between measurements then, the opposing measuring point has moved almost a million and a half miles, or almost one percent of the distance defining the measurement, twice the Earth's orbital distance from the sun of ninety-three million miles.

Failure to take this measurement into consideration totally obliterates any parallax measurement where the most accurate measurement claimed is at least thirty times smaller, less than three thousandths of a percent of a degree of the arc involved.

But, say the sun doesn't move as slow as it is measured to move, say it moves as fast as the Big Bangers say it moves, over forty-three thousand miles per hour. With everything moving away from everything else at that rate, parallax measurements would simply be a joke.

It's lucky that Empirical Science was able to use parallax to determine stellar distances before it was discovered that parallax was useless in determining stellar distances.

Otherwise, we wouldn't know what the stellar distances were!

Robert Hooke wasn't so lucky. His delusion didn't become fact before it was found to be fantasy.

Merry Old England wasn't about to stake its reputation on a chimera!

HOOKE MEASURES SOMETHING

Hooke picked the brightest star in the head of the Dragon Constellation for the parallax measurements needed to prove Copernicus, and thus bring King and country world renown.

He erected a thirty-six foot long perpendicular tube through which he observed the stars passing overhead. However, his results were not readily forthcoming.

"Hooke was exhorted by the President to undertake observations of the parallax of the earth's orb to seconds . . . he was exhorted by the President to make the measurements with all convenient speed . . . Mr. Hooke was ordered during his vacation to observe the parallax of the earth's orb . . . The President of the Royal Society mentioned that he had understood . . ." went the requests.

"Inconvenient weather. . . would be a great indisposition in health . . ." came Hooke's replies.

"He was desired to prosecute carefully this observation . . . Recommended to Hooke to continue to observe . . . He was likewise exhorted to prosecute the observation of the parallax of the earth's orb . . ." the Royal Society minutes continued to direct.

Hooke finally threw in the towel and reported that he had measured the parallax of the bright star in the Dragon's head.

The English astronomer, James Bradley, later finding that the claimed measurement was 30" (thirty seconds!) away from reality, which is to say the error was about six hundred times greater than current claimed parallax measurements, which are in the area of one twentieth of one second, and noting that it was not only grossly off, but off in the opposite direction of stellar aberration, attempted to explain why the measurement appeared reasonable at first. Proffering the explanation that currently supports the entire house of cards that is modern Empirical Science, he stated that he "had no small opinion of the measurements' correctness because the length of the telescope (complexity of the equipment) and the care taken in making them exact (complexity of methodology) were both strong inducements to having thought them so."

In explaining today how stellar distances are computed by parallax measurements, Hooke's delusion is glossed over, much as the venerable Hershel's later discovery of the planet Uranus is, where whatever he saw and recorded was going in the opposite direction Uranus goes.

The end result was that the Copernican proposition remained unproven!

But the measurement of the distance to the stars beckoned. Empirical Science never says die when there is a measurement to find and a ruler that can be built in the mind to find it.

THE MAGNITUDE OF DISTANCE BECOMES "WELL UNDERSTOOD"

Things that are "well understood" in Empirical Science are by necessity divorced from reality. Galileo's inertia was destroyed by the friction of the tides, but it was "well understood" that frictionless movement is not affected by gravity. When Newton proved that gravity was a property of mass, it was "well understood" that the reason those same tides didn't slow the Earth's rotation down was because the mass of the sun and moon literally lifted the water out of the seabeds so that there was no friction resulting from all the sloshing.

And when a couple of shirt-cuff measurements came as close as less than a factor of three, one that found that the stars were a million times further away than the sun, and another that found that the stars were four hundred thousand times further away than the sun, why, it was "well understood" how vast the depths of space were.

Pretty damn vast!

Once something is "well understood" it then takes proof to alter that understanding. No one can prove that frictionless movement is not affected by gravity because no one can produce frictionless movement. Modern frictionless movement, satellites orbiting the Earth, is disproved every day by falling satellites, but Empirical Science claims that satellites fall because the Earth's atmosphere expands and creates friction for the satellites, a nonproveable proposition. No tide on the face of the Earth is predictable by the movement of the moon and the sun, but that isn't important. It is "well understood" that the tides in fact rise and fall with that movement, otherwise the Earth would have long since stopped spinning.

And, it would take extraordinary proof to demonstrate that the stars are not at least one hundred thousand times further away than the sun, proof that can never be forthcoming because there is no way to measure their distances to begin with.

When you can't measure something, the measurements you produce are going to agree with what you already "well understand" rather than what conflicts with that understanding.

All conclusions about the vastness of space were set in stone several centuries before Empirical Science became sophisticated enough to be able to make unquestionable claims as to impossible measurements, in short, before a claim of actual measurement was ever made.

It took two hundred years of a priori conditioning as to the vastness of the distances in space to dictate the conclusions to be drawn from attempting to measure how far the stars are!

THE "WELL UNDERSTOOD" DISTANCES IN SPACE ARE FINALLY CONFIRMED BY PARALLAX MEASUREMENTS

Huygens, of water wave fame, first attempted to establish the distance to the stars by comparing the brightness of the light from Sirius with sunlight and came up with an estimate that Sirius was about seventeen thousand times further away than the sun.

This was dismissed as not far enough, giving an indication of the desire to distance ourselves from the universe, of the need to return to the belief that we are alone and thus simulate the feeling that we are once again at the center of the universe.

Newton's contemporary James Gregory speculated that the sun was a typical star and that Sirius was dimmer than the sun only because of distance, a star's brightness having an inverse ratio to the square root of its distance. Disregarding Huygens' estimate, Gregory recommended studying Sirius when its brightness equaled the brightness of Saturn. Using the comparison of brightness at that point, and knowing the distance of Saturn from the sun, the star's distance could be computed.

Newton did so and estimated the distance of Sirius to be one million times further away than the sun.

James Brady, when not being appalled at Hooke's parallax mismeasurements, concluded that Newton was off by a factor of 2.5, that the stars were actually four hundred thousand times further away than the sun.

Now both these guys were blathering. They didn't have the foggiest idea how far away the stars were, and neither do we, but one of these guys was named Newton, and anything that Newton said, as long as it wasn't that light was made up of particles, was fact.

Galileo had been the first to suggest that because the distance of the Earth from the sun was so great, it could be used to measure the distance to the stars. He proposed using double stars to compute distance. Double stars were stars in the same line of sight so that from the different angles produced by the Earth's orbit, the closer star would move with respect to the further star. (When so many of these optical double were found that it seemed unlikely that they could all be in the same line of sight, unless of course the distances between the stars were not as vast as they were "well understood" to be, it was decided that the majority of optical doubles were actually companion stars, stars that appeared to be close together because they were held together by Newton's fanciful mass/gravity, and once it was "well understood" that these were companion stars, Newton's mass/gravity could be used to compute their mass, how much matter each star contained.)

61Cigni was finally measured in the mid-nineteenth century by the optical double method of parallax measurement to be about eleven light years away, with the angle produced being just over three tenths of a second. This is about one twelve thousandth of a degree of arc, which is an improvement on most parallax measurements, the vast majority of which fall in the area of one seventy-two thousandths of a degree of arc.

SOME REALLY BIG PROBLEMS WITH PARALLAX

A measurement based on the Earth being on opposite sides of its orbit is a measurement based on the Earth being at points that are one hundred and eighty six million miles apart.

This is a pretty big distance. It is pretty seductive to let our minds form a picture of this immeasurably wide base which can then be used to determine how far away stars are. With a big distance like that, how can one think such an undertaking impossible?

The problem is in the execution. While the claim that the ability to measure parallax is based on the sophistication of the equipment used, the simple fact is, even a parallax of three tenths of a second means that in the example of alternately opening and closing our eyes we would be holding our finger over six miles away. While six miles seems like a small figure when compared to the big measurement of one hundred and eighty-six million miles that makes up the base of the Earth's orbit, the point is that the six miles would have a base of three inches, the distance between the pupils of the eyes. If you think it would be possible to discern any movement by alternately opening and closing your eyes with your finger six miles away, even if you could see your finger, then you are a prime candidate for an astronomership, where thoughts become fact without reference to reality.

To put the situation in earthly terms, the claim that parallax measurement can measure the distance to the stars is as credible as a claim that the statue of the bull on Wall St. could be put in place by surveying the tip of one horn using the angle allowed by a hole the size of a dime on the side of the Empire State Building where the location of the Empire State Building is determined by surveying the clouds passing the flame on the torch of the Statue of Liberty through the same hole!

In short, claims that a star's distance can be determined by parallax measurement are chimerical, delusions. Parallax is one of those things that are covered in the first pages of a comprehensive text on astronomy, not to challenge, but merely to acquaint readers to a simple matter of fact, here that finding the distance to the stars is a *fait accompli* and initiates need not ask questions about the underlying assumptions and self-delusion that goes into establishing those distances. No self-respecting astronomer would spend his precious career looking into established facts when Nobels await the empiricist that measures the exact distance to the end of the universe or how much gas must a mass of gas unfurl if a mass of gas must have swirl.

The last part of the facetious example using a hole in the Empire State Building to survey the placement of a bull sixty blocks away, that the location of the Empire State Building has to be determined by surveying the flame of the Statue of Liberty against the clouds floating by in the background, presents an insurmountable problem with parallax measurement, a fatal defect in execution.

Do the finger thing with the eyes, but this time instead of a fixed background, create a movable background with a finger on your other hand. Put your background finger as far away as possible from you eyes, and your foreground finger as close as possible to your eyes. Alternately opening and closing your eyes provides considerable movement of your foreground finger in relation to your background finger.

Now move your background finger up close to your foreground finger and alternately open and close your eyes.

Hardly any movement at all.

Now, put your background finger out away from your eyes again and move your foreground finger next to it and alternately open and close your eyes.

Not much movement here, either.

Place your background finger far away and your foreground finger close. Then move your foreground finger out an inch and alternately open and close your eyes. A lot of movement. Move it out another inch and alternately open and close your eyes. Less movement! Move it still further and still less movement results. Keeping the foreground finger in place and moving the background finger forward produces the same result.

The movement of the finger back and forth as a result of parallax is proportional to the distance *between* the fingers rather than to the distance of the foreground finger from the eyes.

Parallax does not depend on the base, the distance between the eyes. Nor does it depend on the distance of the finger from the eyes.

It depends on the distance between the fingers!

Thus, if measuring parallax with optical doubles tells us anything, it tells us whether the optical doubles have distance between them, although it doesn't give us any quantitative information about that distance.

Measuring distance with optical doubles tells us absolutely nothing about the distance of either of the optical doubles from the base formed by the opposite sides of the Earth's orbit and thus about the distance to the stars!

Why is this fatal defect ignored by Empirical Science?

Knowing the distance to the nearest stars and thus knowing the dimensions of the universe is far too important in comparison with a silly little defect in the ability of parallax to measure the distance to the stars to discard the only tool Empirical Science has to measure those distances!

WHAT IS THE IMPORTANCE OF PARALLAX MEASUREMENTS IN SIZING THE UNIVERSE?

The ancients classified stars on the basis of magnitude, with the brightest stars being of the first magnitude, and dimmer stars trailing off as higher numbered magnitudes resulting in the confusing process of having the dimmer stars' magnitude numbers higher than the numbers of the first magnitude stars. Under the empirical classification system, objects that can hardly be seen through telescopes have magnitudes in the twenties and really bright objects have negative magnitudes, a bit of legerdemain that lends credence to the belief that the measurements are actually measurements.

The ancients quiet naturally thought that the brighter the star was, the greater its magnitude and the closer the star.

Empirical Science is too sophisticated to make such a rash conclusion. A very bright object far away might look the same as a very dim object close-up. Thus, the magnitudes of the stars we see are only apparent. It is not possible to locate stars in space by their apparent magnitude, their relative brightness.

This is an empirically based conclusion.

Because it is theoretically possible that the stars are not ordered by brightness, it's an empirical fact that the stars are not ordered by brightness!

This means that stars of different brightness, and even size and color are strewn haphazardly throughout the universe.

Why would Empirical Science make such a conclusion? Why would it conclude that stars have differing degrees of brightness when there is no basis for making such a conclusion?

The assumption that the distances to the stars was not related to their magnitude creates employment for Empirical Science.

Any old fool could position the stars in the universe by their apparent magnitude, how bright they actually appeared.

Understanding that a star's apparent brightness has nothing to do with its actual brightness, Empirical Science, with its ability to know what can't be known, set about determining what a star's actual brightness, its intrinsic brightness, is so it can place the stars where they actually belong in the universe rather than where they appear to be.

What is a star's intrinsic brightness?

The intrinsic brightness of a star is the actual brightness that the star produces as it burns in space if the empiricist were standing right next to it. It is the result of assuming that stars are not located in accordance with how bright they are. If a bright star is assumed to be further away from the Earth than a dim star, then the further star has a greater intrinsic brightness than the dim star.

Empirical Science has no way of knowing a star's intrinsic brightness, and in fact, intrinsic brightness is simply a concept Empirical Science created to justify the belief that stars are not ordered in the heavens by brightness.

But the only way that a star's distance can be determined is to compare its magnitude with other stars of known distance. Thus, the first task of Empirical Science is to obtain a common measurement of brightness by neutralizing distance. It does this by ascertaining intrinsic brightness and converting it to absolute magnitude.

What is absolute magnitude?

Absolute magnitude is the intrinsic brightness a star would have if it were located 32.6 light years from Earth. If all stars are lined up by absolute magnitude 32.6 light years from Earth, then their relative brightness can be compared.

Why is a point 32.6 light years away from Earth selected?

A circle is divided into 360 degrees, a degree into sixty minutes and a minute into sixty seconds. A second is therefore one thirty-six hundredths of a degree of arc. Because there is no way to know what the distance between the star being measured and the background star is, and thus no way of knowing how much of any measured shift is due to this distance, an imaginary star is created that would produce a parallax shift of one second of arc if the shift were due solely to the distance of the star from one half of the baseline of the Earth's orbit. Because the radius of a circle is equal to just under one sixth of its circumference, called a radian, the number of seconds in a radian can be computed and is equal to two hundred six thousand, two hundred and sixty-five. With the distance to the sun being one second of arc, the distance to the imaginary star would be two hundred six thousand two hundred and sixty-five times ninety-three million, all of these arcane calculations being performed on paper in the confines of the laboratory without getting the lungs exploded, so to speak, in space. In short, the distances to the stars are created with pen and paper.

Look into the telescope and see a little twinkle over a period of six months (complexity of equipment) equal to one six hundred and eight thousand, seven hundred and ninety-fifths of a radian, multiplied by ninety-three million (complexity of methodology) and shazam, Empirical Science has the distance to 61Cigni!

Once the distance to the imaginary star has been empirically determined as a parallax second, or parsec, it can be computed in inches, feet, meters or miles, but because it is so large, it is usually given in light years.

And because it's computable, it's real!

The imaginary star is measured to be 3.26 light years from Earth and 3.26 light years is equal to one parsec. Because this point is closer than the nearest claimed measurement to a star, it is multiplied by ten to get the figure ten parsecs or 32.6 light years, an imaginary point in space used to measure the distances to the stars.

How is the absolute magnitude of a star determined?

When Empirical Science believes that it can directly measure the distance to a star, then the star's apparent brightness is the star's intrinsic brightness. The knowledge of distance and thus intrinsic brightness allows the computation of absolute magnitude because it is known that light diminishes in a precise manner as it expands, with the square of the distance over which it travels. While light diminishes as it expands, Empirical Science views light as being something that exists at the point of observation. Because it has to move stars back and forth to adjust for assumed differences in intrinsic brightness, it expresses the fact that light diminishes inversely with the square of the distance over which it travels in terms of the measured brightness to the observer, that the apparent brightness of a star varies inversely as the square of distance.

Parallax measurements showing a dim star to be closer than a bright star would demonstrate empirically that the far star had a greater intrinsic brightness than the dim star, the empirical portion of the statement being the measurement of the stars' distances. But the intrinsic brightness of the two stars cannot be compared unless they are both placed at the same distance in space. This is where the 32.6 light year line comes into play. Because knowing the distance to a star determines its intrinsic brightness, the inverse square measurement of light can be used to compute the magnitude of the star if it were located 32.6 light years from Earth.

If a bright star was found to be twenty light years from Earth, its magnitude could be increased (reducing its brightness) by a factor of 12.6 light years. If a dim star was found to be ten light years from Earth, its magnitude could be increased by a factor of 22.6 light years.

To translate magnitude into distance or distance into magnitude, Empirical Science uses a magnitude chart which states that each successive magnitude corresponds to the fifth root of one hundred, or 2.512, a little bit of folderol that dazzles the mind with the precision involved in the pursuit of establishing deep space distances giving, as it does, an additional appearance of mathematical accuracy in the process. It also allows for the computation of incredibly vast distances. For instance, if a star's magnitude shift is five, from an absolute magnitude of zero to an apparent magnitude of five, it appears one hundred times dimmer (2.512 to the fifth power). With the star being 32.6 light years away at its absolute magnitude, it's actual distance is 326 light years, the square root of one hundred, ten, times 32.6 light years.

When all of the stars whose intrinsic brightness is known by parallax measurement and whose absolute magnitude can therefore be computed are lined up on the 32.6 light year line, they can each be compared with all other stars whose absolute magnitude is known.

Empirical Science has taken apparent brightness, the way stars appear without knowing how far away the stars are, assumed that apparent brightness does not produce information about comparative distance, convinced itself that it can measure the distance to some of the stars by measuring the angle produced by parallax and then produced a chart of the few stars that are measurable by parallax based on brightness alone, without distance being a factor.

Why has it gone to all this trouble?

Because there are only a handful of stars whose apparent brightness can be converted to intrinsic brightness by believing that parallax measurements exist and determine distance. That leaves billions, nay, trillions of stars whose distances have yet to be determined.

If Empirical Science can use distance to convert apparent magnitude to absolute magnitude, with so many light years equaling so many degrees of magnitude, it can convert a difference in degrees of magnitude into light years in order to determine the distance to a star. For instance, if the absolute magnitude of a star whose apparent magnitude is known by observation can be obtained, then the difference in degrees of magnitude between the stars' absolute magnitude and its apparent magnitude can be computed in light years and added to the known distance of the star at its absolute magnitude of 32.6 light years from Earth.

All measurements in the universe are therefore based on the parallax measurements of close stars!

How does Empirical Science determine the absolute magnitude of stars whose distances it cannot claim to measure and thus whose apparent brightness it cannot convert to intrinsic brightness by knowing its distance?

While Empirical Science can't assume that apparent magnitude, the brightness of a star's light, is related to distance because otherwise it couldn't engage itself in rebrightening, resizing and thus repositioning all the stars in the universe, there's nothing stopping it from assuming that the starlight itself is related to its distance.

DETERMINING INTRINSIC BRIGHTNESS AND ABSOLUTE MAGNITUDE WITHOUT KNOWING DISTANCE

Passing light through a prism results in a rainbow of colors as the distance the light travels through the prism progressively lowers its frequency, and passing very expanded starlight through a prism produces all sorts of distortions in these frequency changes.

Seeing distortions in the spectral analysis of starlight and having a strong desire to compute the distances to the stars quite naturally resulted in the empirical conclusion that a star's spectrum could be used to determine its distance.

But how?

If Empirical Science could find a bunch of stars that were close enough together to relate their intrinsic brightnesses to their spectrums, then the stars could be arranged into a sequence which would allow a determination of the star's intrinsic brightness simply by determining a star's position on the sequence using the analysis of its spectrum.

Once the stars were arranged into a sequence that related intrinsic brightness to a star's spectrum, the star's absolute magnitude could be determined if distance were known and once the absolute magnitude of the stars on the chart were known, then all stars in the universe could be placed 32.6 light years from Earth by placing them on the chart in accordance with their spectral analysis.

Once a star was placed 32.6 light years from Earth, its distance could be computed by the degrees of magnitude between that absolute magnitude and the star's apparent magnitude, how it is observed in space.

It was easy to find stars that were relatively close together. Stars that are close together are called clusters.

But how could the distance to a cluster be determined?

By parallax measurement!

Now, the particular cluster used to establish the star chart that classified the stars by spectral analysis is about forty light years away, beyond the distance at which parallax measurement produces even a color of reliable results. However, the need to create a stellar classification system dictated that the distance to the cluster be measurable, and therefore it was. After all, a group of stars is bigger than a star and thus its brightness more apparent. Empirical Science doesn't quibble about inconsequentials. The purpose is to categorize stars, the requirement is to measure the distance to a cluster of stars, measuring distance by parallax is theoretically possible, it can be accomplished in the mind and it is therefore a fact. Once the parallax of the cluster has been measured and its distance determined, the question of its accuracy never need be reopened, at least by those wishing to put food on the family table by exploring the mysteries of the universe and, in fact, its distance can even be used to show the maximum distance at which parallax measurements have validity.

Once the stars have been ordered into a chart that orders all stars by spectral analysis, then the distance of any star in the universe can be determined simply by analyzing its spectrum and placing it on the chart. If a star's spectral analysis shows it to be such and such a class of star, then its absolute magnitude is absolutely known and its distance is computable by the degrees of difference between that absolute magnitude with its apparent magnitude.

The only difference between this process and the process Huygens used to compare the light of Sirius with the light from the sun is the light from Sirius was the light measured from Sirius rather than the light from Sirius computed as if Sirius was next to the sun at some imaginary point in space, and the light from the sun was actually light, not the darkness of space 32.6

light years away.

Empirical Science has no way of knowing what the spectral distortion of starlight means, or even what it is, it is totally ignorant about what produces the spectrum and has no knowledge of why the spectrum of starlight breaks down unevenly in a prism, and it has no way of knowing the distance to a star, but it creates a system using color, or more precisely, the absence of color, and distance to classify stars, claims that classification arranges stars in accordance with intrinsic brightness, creates a point in space for a star's absolute magnitude, and then uses the conversion of the intrinsic brightness made up from a fantasy based on believing it knows about the star's spectrum and distance to determine the distance itself.

The process is so absurd, it is laughable.

Empirical Science has attempted to produce a formula similar to the time times rate equals distance formula with absolute magnitude and the inverse square measurement of light as the two constants producing distance. But absolute magnitude is dependent on intrinsic brightness and intrinsic brightness can't be known until distance is known. The only distances for which there is a claim to measurability are the distances to the close stars way this side of the 32.6 light year line, and the only stars with actual claims to reliable measurement are the handful of stars supposedly between four and fifteen light years away, and even the claim to measuring these distances is absurd because parallax can only indicate a distance between two stars, not measure the distance *to* the stars, and its use in measuring the distances to the stars has consistently ignored the movement of the base line, the Earth's orbit, in space.

But the point is, parallax measurements are useful because they are the only way Empirical Science can determine the distances to the stars.

WHAT ABOUT THE DISTANCES TO THE ISLAND UNIVERSES, THE GALAXIES LYING OUTSIDE OUR OWN GALAXY?

A star, Cephei, can be seen to vary in brightness over time. We can't watch the star one hundred percent of the time, but the mind works to fill in the blanks in the pictures it forms, and when it sees a star vary in brightness, it attempts to put order to the rate the star varies.

Think about it, because astronomers can't. You see a star from night to night if the skies aren't covered with clouds. The task is to determine the regular cycle of a star that appears to vary in intensity. It grows bright, and then it grows dim, and then it grows bright again, then it grows dim again. You see the star maybe five nights out of ten, one or two of which, if you're lucky, might be consecutive, and then only one or two months out of a year and never from the same angle twice in any one year.

How do you determine its regular period?

More to the point, why would anyone even consider trying to determine its regular period?

Why would anyone think it *has* a regular period?

These considerations aside, Cephei has a regular period of 5.4 days during which it rapidly achieves its maximum brightness and then slowly recedes into its minimum brightness, the variation in brightness being on the order of one magnitude.

What in the world does this tell us about distance?

To know distance, we have to know intrinsic brightness, for it is by knowing intrinsic brightness that Empirical Science can place a star by its absolute magnitude on the 32.6 light year line, and thus compute a star's distance from Earth.

How could the fact that a star changes magnitude allow us to determine the star's intrinsic brightness?

Empirical Science determines intrinsic brightness of the stars close to the Earth by comparing their imagined distances. It determines intrinsic brightness by all other stars that can be resolved by comparing their spectrums.

Thus, determining intrinsic brightness is simply a matter of finding something to compare starlight with. If stars had pigtails, their intrinsic brightness would be determined by comparing oinks.

There are two things going on with a variable star: It is changing magnitude, which, while not intrinsic brightness is still brightness. And it has a period.

Thus, there is something, the cepheid's period, to which the brightness can be compared.

Why exactly does Empirical Science want to believe that the length of a variable cepheid's period determines its intrinsic brightness? After all, it can determine a star's intrinsic brightness by lining the star up on the star chart according to its spectral class.

Galaxies, islands of stars far, far away in the blackness of space, don't readily give up the spectral fingerprint of their individual stars.

Without a spectral fingerprint Empirical Science can't classify the star and thus fool itself into thinking it knows the distance to the galaxy. What it has to do is find lighthouses in the night, beacons in the void, cepheid variables in neighboring galaxies blinking on and off to fool itself into thinking it knows the distance to the galaxy. If all cepheid variables with identical periods have the same intrinsic brightness, all Empirical Science has to do is measure the period and it can calculate the intrinsic brightness.

And of course, with the basic tenet of Empirical Science, what you make up in the mind you will always find, firmly behind the proof that the intrinsic brightness of variable stars is directly proportional to their periods, variable stars were found in neighboring galaxies and their distances determined down to the last fraction of a parsec.

Where's there a whim, there's a way!

PROVING THAT THE PERIOD OF VARIABLE CEPHEIDS CAN BE USED TO DETERMINE THEIR INTRINSIC BRIGHTNESS

In order to prove that the intrinsic brightness of variable cepheids was proportional to their periods, it was necessary to line them up by absolute magnitude on the 32.6 light year line.

The first question one might ask is how can something that grows bright and dim and bright again have an absolute magnitude. While astronomers give a lot of idiot explanations for why a star's brightness varies, the simple explanation is the same explanation for why things burn at a varying rate on Earth, they are burning at times faster than the fuel would normally permit because of changes in their environment. With a variable star this occurs as a result of the eccentric orbits of hot planets. Moving two hot objects closer together causes each to heat up as a result of its proximity to the other. A hot planet moving closer to a star causes the star to flare up, much as the conjunction of planets in the solar system produces solar flares on the sun.

But anything that is burning irregularly has no intrinsic brightness and absolutely no absolute magnitude.

Not to worry. As long as absolute magnitude is calculated at the same point in the cycle for all stars, empirical purity has been maintained.

So the spectral class of variable cepheids is taken off the star charts, their intrinsic brightness determined, their place in space adjusted so that they are 32.6 light years away and their periods compared.

Lo and behold, their periods correlate exactly with their intrinsic brightness!

Any cepheid that doesn't correlate can have its intrinsic brightness adjusted by the creative use of space dust. If the period is too long to agree with the proposition, then the star is dimmer than it should be because there is heretofore unnoticed space dust in the way and if the period is too short, why, then, too much space dust has been accounted for and some needs to be removed.

Because intrinsic brightness dictates absolute magnitude, any period can be made to coincide with any absolute magnitude simply by altering the cepheid's intrinsic brightness by adjusting the amount of space dust that might exist between the cepheid and the telescope.

If anyone thinks Empirical Science actually went through the above exercise, he isn't acquainted with the ways of Empirical Science. Why go through the motions when the outcome is known? If it is "well understood" that all cepheidic periods can be made to correlate to intrinsic brightness, if something is theoretically possible, if it is measurable in the mind, then it exists in reality.

It is so "well understood", in fact, that the Amateur Astronomer's Handbook points out that cepheids "have been closely studied and offer little scope for amateur work." In other words, measure "well understood" facts and risk being ignored. A lot of amateur astronomers apparently risk being ignored if the American Association of Variable Star Observers is any indication. There is always a lot of amateur activity in the two areas that don't follow Empirical Science's predictive delusions, variable cepheids and asteroids. Because of the difficulty inherent in viewing variable stars from a single vantage point, observers all over the globe view the apparent magnitude of variable stars and report the results to the Association. The resulting graphs are anything but regular, and in a manner that makes correction by space dust impossible, removing any basis for using the period of variable cepheids to determine intrinsic brightness.

However, without the proposition that the variable cepheids' periods vary directly with their intrinsic brightness, it would have been impossible to prove that the Andromeda "nebula" was actually a galaxy just like our own, and outside the boundaries of our own galaxy.

And that was what the proposition was made up to prove in the first place!

WHEN VARIABLE PERIODS AREN'T ENOUGH

When the distances computed by Empirical Science become so vast that even its most incredulous empiricists can't stomach the claim that variable periods determine intrinsic brightness, Empirical Science has to develop other yardsticks to dazzle the senses and confuse the mind.

At distances where intrinsic brightness is no longer computable by variability, variability has to be abandoned for other pigtails.

The pigtails selected were clouds of ionized gas. Empirical Science has absolutely no physical description of what is going on when what is measured on Earth as the ionization of a gas occurs, which is all the more reason why it can point at bright spots in the night sky and say, there rests a blob of ionized gas. Comparing the brightness of ionized gas clouds found in different galaxies therefore allows the determination of the intrinsic brightness of galaxies (intrinsic brightness of a group of stars!) and thus the computation of the distance to galaxies at the end of the universe.

Of course brightness is the quality the ancients used to order the position of the stars in the universe to begin with, the quality Empirical Science chose to ignore in favor of using parallax measurements to determine the imagined intrinsic brightness, and thus provides no empirical information as to distance.

But, hey, parallax measurements don't work at these distances so Empirical Science has to use what it can to come up with precise measurements for the distances to the stars and galaxies. And why stop at comparing ionized gas clouds?

If Empirical Science can get this far with made-up up stuff, why not go whole hog?

USING AN EFFECT AS ITS CAUSE

Knowing absolutely nothing about the nature of light, how it was produced, how it moved, what happened with it once it got to where it was going, it was very easy for Empirical Science to say that light was the same as sound. As sound waves traveling through the medium of the atmosphere can be elongated or compressed, so can light waves traveling through the emptiness of space.

That settled, having no idea whatsoever about what happens when light goes through a prism, being totally ignorant of the nature of color, acting like monkeys when viewing light refracted by a prism (monkey sees color, monkey says order), it was very easy for Empirical Science to say, compressed light waves look blue and elongated light waves look red. If a train is moving away from us, its sound waves are elongated and their pitch changes. If a star is moving away from us, its light waves are elongated and the spectrum it produces moves measurably into the red.

This pile of donkey doo results in the conclusion that everything in the universe is moving away from everything else, and the further away it is moving away, the faster it is moving away, truly one of the most ludicrous fantasies ever created by the mindless musings of man.

But, if something is moving, and it is moving at a certain rate, the old time times rate equals distance formula actually does become feasible. By golly, Empirical Science should be able to use this empirical fact to calculate the empirical size of the universe and the relative distances of all matter within it.

The proportionality of the red shift to the rate of expansion then becomes Hubble's Constant and once Hubble's Constant is known, Empirical Science can use the spectral analysis of starlight to position any star or galaxy in the universe, locate its speed, direction, distance, and what it is having for dinner (the fuel it's consuming).

Empirical Science knows what Hubble's Constant does. As it drops, the distances it calculates increase proportionally. With the Constant dropping by a factor of seven since its creation early in the century, the universe has been turned into one very big, very rapidly expanding bubble.

The only problem is determining what Hubble's Constant is!

The way Hubble's Constant is determined is to use the distance ladder, going from parallax measurements to spectral classification to cepheid variables and ending up with ionized gas clouds, all the while manufacturing and eliminating space dust wherever necessary or unnecessary to make everything come out just so (the size of the bubble and its rate of expansion have to equal the age of the universe, all empirically calculable quantities for sure).

But Empirical Science can't be too accurate when it's determining the size of the endless expanse of nothingness we call space and fondly refer to as the universe, so as a service to mankind, it forever hones its measuring tools.

Using the clarity and contrast in the arms of spiral galaxies as a comparison tool instead of ionized gas, Hubble's Constant was found to be a little bit off kilter.

And when it was discovered that comparing the rotational velocity of galactic disks could also be used to determine intrinsic brightness, the distance ladder came full circle.

The rotational velocity of a galactic disk was determined by the red shift whose coefficient was Hubble's Constant which was what was being changed by altering intrinsic brightness by rotational velocity!

As Hubble's Constant increased the size of the universe, it also increased the number of stars in the galaxies.

But one thing didn't increase.

The distances to the nearest stars, determined by parallax measurement, the distances on which Hubble's Constant was based, the first step on the delusional distance ladder remained the same.

But then that distance can't be determined to start with!

Ye who allow Empirical Science to enter abandon all pretense to reason because sense is no more.

Empirical Science determines the distance to the stars by determining that the distance to the stars is equal to the distance to the stars and is therefore precisely the distance to the stars.

SO HOW FAR *ARE* THE STARS?

In "Mathematics," Chapter Seven of *At the Gates of the Citadel*, the first volume of The Copernican Series, a little spoof is perpetrated. In showing how to calculate the speed and distance of the sun, it is posited that if current methods of computing distance in space produce a sun going about forty-three thousand miles per hour because of overestimated distances, and the sun is actually going about three hundred and twelve miles per hour, then we should be able to take the fraction produced by putting the sun's actual speed over the miscomputed speed, apply it to miscomputed distances and come up with the actual distances in space. This little bit of tongue-in-cheek has been taken seriously, which shows how far gone Empirical Science is. The reason that the fraction won't produce an accurate estimate is that the bottom of the fraction, the speed of the sun produced by Empirical Science, has no basis in reality, and any distance to which the fraction is applied is outright fantasy.

We have no idea how far it is to the nearest stars.

We can, however, make a reasonable estimate.

One of the space probes recently sent back pictures of the sun from far planetary orbits, and these pictures showed a very bright star. Double or triple the distance and we would have a star that would be equal in magnitude to some of the stars we see from Earth, demonstrating in reality that our sun would appear as a star at only two or three times the distance to Pluto.

Another approach to estimate how far the stars are is to ask how close the stars could be together.

All motion in the solar system results from what goes on in the sun's emission field as it expands out from the sun and back away from the direction of the sun's rotation. As the sun's emission field becomes weaker, diminishing as it does with the square of distance, the interaction of the planets within that field becomes less. One of the interactions, the rate at which a planet falls in the sun's emission field, becomes less with distance. The interaction of a planet's emission field with the sun's emission field also determines the planet's rotation and thus its axis.

Uranus lies on its side, attesting to the weakness of the sun's emission field at that distance. Neptune's orbit hasn't been measured long enough to establish and it's questionable whether Pluto even has an orbit, further attesting to the weakness of the sun's emission field at these outermost distances.

Thus, by Pluto's orbit at least, the effect of the sun's emission field is extremely weak. Triple the distance to Pluto and a star the size of our sun would look like a star, and could exist without attempting to orbit our sun, or without our sun attempting to orbit the star.

Add another distance to Pluto and we would have a reasonable estimate of the average distance between single stars in the bright centers of galaxies.

We are in one of the spiral arms of our galaxy, and the stars in the arm are drifting apart, increasing the interstellar distances between them so that a viable estimate for the closest stars would be in the neighborhood of six to eight times the distance of Pluto from the sun. (The distance of the closest stars computed by the facetious fraction in Chapter Seven of *At the Gates of the Citadel* is sixty-six times the distance to Pluto, about one sixth Huygens' estimate derived from comparing Sirius to the sun.)

If these distances seem minuscule, their reality is bolstered by an absolute limitation on the distances to the stars that we see. Light from stars expands over the surface of an expanding sphere and as it does so, it diminishes inversely with the square of the distance over which it travels.

It eventually expands out of existence!

We can't see light that has expanded out of existence no matter how strong our telescopes, and no matter how sophisticated our electronic detectors.

The New Astronomy thinks it can because otherwise there would be no New Astronomy.

But what we see is limited by the fact that light diminishes inversely with the square of the distance over which it travels and that limit places stars a whole lot closer than they are "well understood" to be.

Galaxies don't have billions of stars, they have tens of thousands of stars. They aren't a hundred million light years across, they are trillions of miles across.

Stars and galaxies are not separated by endless oceans of space, they are right next door.

In short, if we can see it, we can probably get to it!

8. THE EARTH IS BILLIONS OF YEARS OLD

CERTAIN THINGS ARE NOT KNOWABLE

We want to know certain things about new acquaintances. One of the first things we want to know is how old someone is. Things in our existence have beginnings, middles and ends. Because we have beginnings and middles and ends, we look for beginnings, middles and ends in everything around us.

Several thousand years ago, when there were no records, finding out how old someone was, when their beginning was, could present a problem. Without birth certificates, and without an accurate calendar, we could only estimate age. The four ages of man were the rough periods of a person's course through life, and putting a person in a particular age involved looking at the various characteristics and abilities, actions and behaviors that made that person up. Does the person crawl, is he vigorous, does he appear young, does he look wise, is he losing his facilities, is he wrinkled, does he ramble, these are characteristics that can be used as indicators of age. But without a record, we have no way of knowing the actual age.

Today, we keep fairly accurate records. As a result, although a lot of cemetery shoppers pick up aliases off tombstones, we can know with some degree of certainty how old someone is.

The age of a new acquaintance is, therefore, probably knowable. It is something that we can find out. It is a fact that can be established as well as actual facts in physical reality, facts such as a person's height, weight, eye or hair color, even sex.

What makes the question of age more ambiguous than actual facts is the actual fact that the person's age depends on something that happened in the past and the past no longer exists!

Because the past no longer exists, it is out of our reach and without accurate records or without records upon whose accuracy we can place reliance, we would have to fall back on estimating age from characteristics.

With records, age is knowable. Without records, age is approximately knowable.

Why is age approximately knowable without records?

Because we have a population that we can look at, we can see people in various stages of aging, we can look at records that provide the characteristics that occur throughout the aging process and we can then compare this information with the characteristics of the person whose age we are trying to estimate.

Age is approximately knowable without records because we have experience which we can apply to the facts to estimate the age.

But there are some things that are simply not knowable because we have no records, and we have no experience that provides characteristics with which comparisons can be made.

We deal with beginnings and ends because we have beginnings and ends, but we have no way of knowing what occurs before the beginning or after the end. We don't know where we were before we were born or where we go after we die because neither are current facts, and we have no experience we can use to bring to bear on estimating, on producing the knowing.

When we can't know something, then we simply can't know it.

As a result, we make things up!

There is nothing that distinguishes making up things that deal with nonexistence from making up things dealing with existence, reality, when the facts about that reality are out of reach, no longer exist.

However, we deal differently with knowing about the unknowable where there is no physical existence for the unknowable, questions of where we were before we were born or where we will go after we die, and where there is physical existence.

No one would seriously consider statements about an afterlife to be statements dealing with physical reality, and all would agree that such statements belong in the realm of the occult, are mystical in nature, provide the basis for religious systems.

But there are many questions about physical reality the answers to which we cannot know. We can't know what's in the center of the Earth or the sun because it is out of our reach. We can't know how far away a star is because we can't put a measuring rod against the distance. We can't know what makes up the atom because atoms make up our measuring rods. We can't know how big or how small the universe is or when it began or when it will end.

And we can't know how old the Earth is!

When we can't know something, we endanger ourselves if we pretend we can.

It is one thing to be mistaken about something, to think we know something when we don't. The results can be catastrophic, but at least the delusions that lead to our demise will be honest delusions, the result of simple ignorance.

It is quite another thing to pretend to know something that we cannot know.

The results are the same, of course, catastrophic, but in this case our demise can be attributable to simple stupidity.

Ignorance is unavoidable, stupidity inexcusable.

Claiming to know something that we can't know is shamanism and those that lay claim to unknowable knowledge are shamans.

Walking around with the delusion that committing suicide will transport us to a space ship in the tail of a comet is a private affair and believing that the sun moves around the Earth and the Earth is thus at the center of the universe is a mistake of fact born of ignorance and arrogance.

But arrogating knowledge where knowledge is clearly not possible is criminal, an offense against the paying public that puts faith in those it charges with obtaining actual answers about questions dealing with physical reality.

PRETENDING TO KNOW WHAT WE CAN'T KNOW

On the morning of May 1, 1902, in the city of St. Pierre, Island of Martinique in the West Indies, the rumbling continued apace. The night had seen a continuation of the bright flashes of lightening that had been turning the night sky into spectacular bursts of daylight for the past week. Puffs of smoke and ash continued to rise above Mount Pelée as they had done for the last twenty-three days.

The thirty thousand residents of St. Pierre went about their business as usual because those charged with knowing claimed that St. Pierre had always existed and it would always exist, and the activities being carried on by the neighboring volcano were of no concern because they had never been of concern before.

On the morning of May 2, 1902, the city of St. Pierre, Island of Martinique in the West Indies ceased to exist. The thirty thousand residents that had gone about their business on the morning of May 1 were now simply gone, no longer in existence.

The pictures we form of physical reality, the answers we make up for questions the answers to which we cannot know, dictate our survival. How we go about answering questions for which there are no answers has been the basis for some of the more profitable movies of the last decades. *Jaws* dealt with a known danger, a shark, that struck intermittently. A lone crusader tried to alert the townsfolk to a threat they were downplaying because it was against their commercial interests to recognize the threat. The shark in *Jaws* was the volcano in *Dante's Peak.*

How one accepts the danger may well determine whether one survives or not. We can move and nothing will happen, or we can move and something will happen, or we can stay in place and nothing will happen, or we can hit the jackpot, stay in place and be wiped out.

Personal decisions whether to stay in St. Pierre or swim with the sharks pale when the question becomes one of predicting catastrophes, where a hurricane will strike or when an earthquake will occur. Moving a million people away from a shore or out of a city could be expensive, and if nothing happens, embarrassing. As much as we want to believe that we can know where hurricanes will land or when earthquakes will strike, these things are unknowable. It's a thankless task to predict wrong, and irresistible not to predict at all.

But there is one area where predictions about what course events will take is not a prediction, but a fact in front of our faces.

The sun is burning furiously in the sky, and as it does so, it is getting measurably smaller. It is, of course, large enough to continue to get smaller for some time to come, but the fact is, whatever is fueling the sun is being used up as it is burning.

Next to the sun is a dried out hunk of rock, Mercury, that orbits in an elliptical orbit that causes it to alternately heat up and cool off as it approaches and moves away from the burning sun. Next to Mercury, moving in a nearly circular orbit, is Venus. Venus, however, doesn't heat up and cool off as it moves around the burning sun. It is hot as hell all by itself, hotter at its coolest place than the Earth is at its hottest.

The Earth, of course, is the ideal, with polar ice caps, toasty equatorial regions, and everything else in-between, but it's moon is as dead as Mercury. Because it is further from the sun, however, the moon is as cold as the night of space.

And the night of space is cold.

Very cold!

So it is only reasonable to come to the conclusion that the sun and the planets that orbit it, and the moons that orbit the planets formed at some time in the past hot, and that because they occupy cold, empty space, they have been cooling off ever since, with the sun, because of its large size, red hot, molten on the surface, with the matter that is burning not yet congealed enough to present a solid surface to the coldness of space, with Venus, much smaller, it's surface congealing, producing a haze of light elements that appear as clouds to buffer the still molten surface from the emptiness of space, with the Earth cooled off enough for the clouds of light elements to have recombined into water that can exist in pools on the cooled rock of the surface and provide the basis for life, and the moon, smaller still, too cold to support life or hold water to its surface.

While a planet losing its water opposes the concept of gravity as a property of mass and the idea that Venus is hot because of internal processes that dictate it be a late comer to the solar system opposes the concept of a solar system set in motion some time in the past by a historical force and maintained in that motion unchanged, undisturbed to the present day, the picture presented of planets formed hot and cooling down in the coldness of space is not only viable, it is inescapable. It's awfully hard to look at the sun and say that it isn't burning, that it isn't consuming what it is burning, growing cold as it does so, it's awfully hard to look at Venus and say that it isn't cooling but rather growing hot in its circular orbit because of sunlight, and it's darn near impossible to look at the moon and say it's not what it is, a hunk of cold rock.

So why do we look at the Earth like the St. Pierrens looked at their rumbling volcano, what was yesterday will be tomorrow, when we know that heat flows from hot objects when hot objects occupy cold places? How is it that we can maintain the absurdity that the moon and the Earth were formed out of a swirling mass of gas at the same time, and that the moon was formed cold and the Earth rapidly cooled down to its present temperature where it has remained ever since?

How can we not see a progression from the sun to the moon with the Earth at some point in-between?

We pretend to the knowledge of the Earth's age and condition during that age, that the Earth has maintained its temperature for some four billion years, because of a mistake in the concept of evolution, that evolution is Darwinian, it involves the mutation of one species to the next rather than the evolution of characteristics which can be transmitted for use as needed as species evolve to match the existing environment.

The idea that dinosaurs morphed into birds and as a result required millions and millions of years to lose the weight and gain the wings needed to soar through the air controlled how we made decisions, created processes and procedures which would make us believe we had answers to questions for which no answers were possible.

We have constructed a picture of our future on Earth, an endless sojourn on a planet that will never cool off, in fact a planet whose only danger is exactly the opposite, the danger that it will heat up and fry us all, in the same way that the Gilchrists construct the Nancy cartoons. In a Nancy cartoon, the last panel is drawn first and then all prior panels are drawn to give the final panel meaning.

In determining the unknowable age of the Earth, the final panel, a timeless past in which all forms of chance experimentation could bring forth complex life forms, is constructed, and then all prior panels, realistic estimates of the Earth's age made by Lord Kelvin based on its rate of cooling, the origin of life requiring some predecessor, if not an egg or seed, the invisible spore, mountain ranges accreted by magma buildup, an eon and more of Mount Pelées when seashells are found on Himalayan ridges and other mountain ranges which contain unweathered and therefore young peaks, sedimentary deposits containing an orderly sequence of fossil remains so that we can pretend to having an actual "record," a kind of cosmic courthouse containing species birth certificates, and radiometric dating, a process that produces a sort of replicating energy, radiation, that by continually heating up the cooling Earth, resets the atomic clock it purports to produce, are all drawn to support, to give meaning to the picture of species evolution presented in the final panel.

ATTEMPTING TO KNOW THE UNKNOWABLE RESULTS IN BELIEF

The age of the Earth is a fact. The Earth is here, it exists, it had to come from somewhere, and it had to come from somewhere at a specific time and that time is in the past, so the age of the Earth is a fact.

It is just an unknowable fact.

Any explanation for an unknowable fact cannot itself be fact, it must forever be a product of the mind, a made up fantasy, something for which belief is accorded.

When we are dealing with beliefs, we are not dealing with reality. As a result, we tend to claim our beliefs facts and then undertake crusades to eradicate any opposing beliefs.

Speculations about the age of the Earth are so intertwined with religious beliefs that the controversies that surround the question will, in all probability, survive the twentieth century and plague the foreseeable future.

The best phrase with which to summarize the controversy, or at least its tone, is the assertion that the Earth is flat.

"Flatearther" was a term manufactured by Darwinists to avoid the religious issues inherent in opposing evolution.

No one at any point in the history of the planet ever made the claim that the Earth was flat!

Darwinian supporters accused their opponents of believing that the Earth was flat, implying that at some point in the past, when science was fantasy and people held beliefs not supported by fact, ignorant savages held the belief that the Earth was flat. Opponents of evolution are thus flatearthers, ignorant primitives. Taking any other tact would result in a challenge to the Church's teachings dealing with creation. If man was created by God, then he certainly didn't evolve from simpler life forms, in the words of the anti-Darwinists, apes.

Calling their opponents "Godlovers" didn't seem to have the necessary sting to it and involved a bit of danger, the least of which was public scorn, so the Darwinists created a belief that ignorant people believed at one time that the Earth was flat and that, by implication, also believed that God created man and that these flatearthers were not dealing with the facts of reality, the science that the Darwinists were making up to explain the origin of life on Earth.

In many respects, the argument goes to the very core of a science that claims that its beliefs are based on fact. All of Modern Science's claims, that gravity is a property of mass, that the Earth rotates as a result of angular momentum, that planetary orbiting is the result of the solar system condensing out of a swirling mass of gas, that the universe was created in a very large explosion, that light is a wave, electricity a moving charge and magnetism the result of molecular magnets, obtain their authority from the belief that they address the facts in reality because they purport to verify those facts by measurement as opposed to the religious creation myths, which claim their authority from revelation.

Actually, the difference is one of degree. Creation myths are designed to explain how things got the way they are, and scientific myths dealing with creation merely contain the added claim that the way things are have been verified by measurement. There is no way to measure the formation of the solar system, either as to how or when, and there is no way to measure how species evolved, so that any difference in explanation is only as valid as the name-calling opportunities presented to each, "Your mommy was an ape!" or "You're an ignorant flatearther!"

No matter who says what with how much authority, we're still dealing in beliefs, made-up stuff the validity of which will never be known because the facts they deal with, the source of current existence, no longer exist and are therefore unknowable.

CLOCKING THE AGE OF THE EARTH

The first modern attempt to determine the age of the Earth occurred in the seventeenth century, several centuries before Darwin. The bible starts off with a bunch of begats, allowing the construction of a genealogy of sorts all the way back to Adam and Eve in the Garden of Eden. By putting an average life span to all of the begats, Bishop Ussher placed the date of creation on October 22, 4004 years before the birth of Christ.

This date is clearly based on religious authority. A hundred years before, Copernicus had published his observation that the Earth, rather than being the center of the universe, moved around the sun. When the Earth was the center of the universe, it didn't have to come from anywhere. But as people became accustomed to the idea that the Earth was like a rock hurtling around the sun, the question of where the rock came from, and when, came into play. Religious authority abhors unanswered questions (as does scientific authority) and when questions are asked, answers have to be generated, otherwise what kind of authority would it be? No one accords authority to institutions that go around saying, "I don't know."

Authority is accorded to anything that has the power to pretend authority!

Just as the flatearth argument allowed Darwinists to attack religious creation claims without mentioning God, so did the good Bishop's absurdly specific claim give Darwinists an endless opportunity to attack biblical creationism without mentioning the Bible. The world was still in the throws of updating its calendars to accommodate itself to errors in computation that had put it almost two weeks into the past so that the stupidity of even attempting to claim a specific date thousands of years in the past was a joke on its face.

Claiming the superiority of your own beliefs by denigrating the beliefs of your opponents, Bishop Ussher's claim is the opening gun professors use to teach evolution to this day, is certainly a noble undertaking of longstanding practice, but it adds not a whit to the validity of your own beliefs, and only a fool would believe that it does.

However, asserting that Modern Science is an intricate web of belief systems with no basis in reality other than the claim that the beliefs arise from precisely measuring reality invites the rejoinder that emphasizing the belief nature at the basis of Modern Science overlooks the way in which the probability of a conclusion supported by many interlocking and mutually reinforcing clues can be close to certainty, even if every clue by itself is less than definitive. Modern Science bases its authority on the overriding belief that the existence of many different strands of evidence can transform a hypothesis into a fact without regard to the validity of any particular piece of the evidence.

In other words, Modern Science overcomes the claim that it is based on belief by claiming that the careful construction of all panels of the comic strip turns the final panel into a fact!

THROWING OUT THE BABY WITH THE BATH WATER

Three specific belief systems mark recorded history in the western world. In the first, Aristotle set forth a comprehensive scheme for the existence of matter and motion. The central feature of that system was that the Earth was the center of the universe around which all objects in the heavens moved. The second belief system, Christianity, was sourced in the old testament creation myth and after a thousand years itself incorporated, or rather accommodated Aristotelian concepts of matter and motion into its creation myths.

When Copernicus described a solar system in which the Earth was not the center of creation, the Aristotelian idea that the Earth was the center of the universe was so tied up with the religious concept of man as the center of creation that to oppose one, the physical relationship, was to oppose the other, the ego based religious precept of man as central to the universe.

At the time, education was church based, Aristotelians were called Philosophers, and the entire universe was viewed through the lens of the church-based Aristotelian authority. Because the controversy over whether the sun went around the Earth or the Earth around the sun could not be settled by measurement, a scientific belief system arose as the third belief system, a system whose sole goal was to use physical measurements to replace the Aristotelian belief system incorporated into church teachings.

You don't throw out one part of a belief system and keep the other parts. The Church may well have had to accommodate Aristotle because physical reality was fact and many revelations did not address actual facts in physical reality, nor would it be desirable for them to do so where revelation opposed those physical facts, but the group that sought to destroy the old way was claiming only measurable fact to start with, was claiming that answers could be obtained without revelation, without subjective belief (such a claim itself being a belief).

As a result, anything the believers in an Earth centered universe believed in was as wrong as the belief in the Earth centered universe.

The facts set forth in the Bible were just as valid as the facts that led Bishop Ussher to come up with a precise October date for the day of creation.

They weren't!

Not only were the facts in the Bible not valid, they were a source of amusement, a subject for derision. What was in many cases historical record-keeping in biblical format was categorized as revelation and put in the dustbin for belief not worthy of scientific consideration.

One of the facts set forth in the Old Testament dealt with the Great Flood.

The flood story is not specific to western thought or the Old Testament. It is universal. Every religion and cultural tradition on the face of the planet has a flood story, with groups of people on every speck of land on the globe recounting tales of their past that includes the flood story. At last count, there have been over five hundred separate flood stories recorded, all essentially identical.

But for the western scientific belief system, the flood story was a part of the Bible and therefore not only myth, made up fantasy, but in addition, something that absolutely did *not* happen!

The flood was a catastrophe that happened suddenly, and in doing so, changed the face of the Earth and the nature of the life forms occupying it. Church-based Aristotelian precepts were not just wrong, they described the opposite of what occurred. If the Bible described a catastrophe, the catastrophe didn't just not happen, no catastrophe ever happened.

Thus, any idea that catastrophic events occurred in the Earth's past had to be thrown out with the idea that the sun went around the Earth. The young effort to determine how to view the Earth's history was treated like a Ming Dynasty courtesan's foot, bound and crippled, with the result as slavishly adored!

By rejecting a catastrophic view of the Earth's history because the biggest catastrophe possible, a worldwide deluge, was related in the Bible, the seekers of the new truth, the anti-Aristotelians, were left with the opposite of catastrophe, sudden change, uniformity, no change.

MOUNTAIN RANGES AND LAYERS OF SEDIMENT

The evidence that mountain ranges were thrust up in catastrophic events litters the mountain ranges themselves, with sea level fossils littering mountain sides, remnants of tropical plants necklacing mountain peaks which are themselves unweathered and therefore of recent origin and the geologic structures themselves clearly the result of upthrust rather than accretion.

The evidence that mountains were built up gradually by magma extrusions is nonexistent. However, because magma comes out of volcanoes and builds up the sides of those volcanoes, it is conceivable that very big mountains could be built up by magma flows and therefore they were. In a modern version of the extrusion belief, instead of the magma pouring out of a hole in the ground, it lubricates the movement of giant plates which float around on the surface of the Earth, occasionally undercutting one another, pushing up mountain ranges in the process.

There is no evidence that plates are floating around on molten magma, but because it is possible that they could be, this is the explanation for mountains, with perhaps the exception of the Rocky Mountains, the Atlas Mountains, the Andes and chains in the Antarctic to name a few.

When these mountains in turn contain evidence that supports the biblical flood which could never have taken place because it is a religious myth, then immense, slow moving ice ages, acting over eons, were created, ice ages whose cause defies explanation, but which are needed simply to explain how giant boulders, hundreds, sometimes thousands of tons of solid rock called erratics because they are out of place, could be transported vast distances from their geographical sources, some found on ground higher than ice soars, others in the middles of desserts where no ice ever tread.

Without a catastrophic occurrence, the physical evidence that catastrophic acts had to have wrought has to be explained away by something else because the catastrophic acts could never have happened.

Better to have an unexplainable ice age than a biblically described event!

The basis for all of these self-delusions, these beliefs in things which don't exist, magma extruded mountain ranges, roaming continental plates, ice ages, is the need to know that which cannot be known, the age of the Earth.

Lyell observed in the mid-nineteenth century that layers of sediment laid down layers of rock one atop the other. This stratification meant that each layer represented a geologic era in the history of the Earth. Because some of these strata contained the remains of the life that existed when the strata was laid down, the strata could be used as a clock to gauge the age of the Earth and the order of development of life on the Earth.

There were two assumptions that had to be made if strata were going to be used to write the history of the Earth. The first was that the thickness of the strata could be converted into years, that the time each strata represented was knowable.

This, of course, was simple delusion.

The second assumption required to turn the observation of strata into a history book, a "record," the word chosen because it produces the illusion that beliefs are facts, was that the strata had to have been uniformly laid down over the entire face of the Earth throughout the eons. It wouldn't do for the English, in whose country the idea was hatched, to build one clock based on a reading of their strata, and the Americans to build another clock using the strata in North America. It was necessary for all strata to represent the same history no matter where on the face of the Earth the strata appeared.

This is a complex delusion indeed!

If the Earth had been mucked up by a bunch of catastrophes, if floods had poured throughout the land, mountain ranges had been thrust up here and there in response to earthshaking events or monumental earthquakes had mixed and mismatched the strata, then the clocking device of the strata would be useless in telling us that which we could not know.

As knowing is much more pleasant than not knowing, catastrophes had to be done away with. Because they were biblically excluded to start with, anything in the Bible was bunk, a doctrine could be created and believed in which excluded the possibility of catastrophes mucking up the stratigrapical time clock. To underline the need for absolute belief, unthinking acceptance by the faithful, the doctrine was given an extremely religious sounding name, uniformitarianism, the uncapitalized "u" identifying it as a nonreligious belief. This belief, that all geological phenomena are the result of gradual forces that have operated uniformly from the origin of the Earth to the present time, is sacrosanct today, surviving even the occasional dinosaur death-dealing asteroid.

EVOLUTION DICTATES THE CLOCK'S RATE

Attempts to come to grips with some sort of explanation for life on Earth that didn't include the active hand of a metaphysical creator had been swirling around for hundreds of years before Darwin came up with his narrative on the evolution of species.

Because Naturalists were Empiricists whose purpose was to use measurement systems to classify life, the various forms of life on Earth were in the process of being categorized, placed into neat compartments so that when people were talking about something, they knew what they were talking about. "Species" is the category defined by the ability to interbreed. Darwin's story dealt with the formation of life at the simplest level and its assumed evolution into more complex forms. Because interbreeding was a natural barrier to the transition of one species into the next, Darwin was concerned with the evolution of species.

Because species evolution is a formidable mental jump as well as an impossible physical transition, the "how" was left to chance, species thought to evolve even today from one to the next as a result of happenstance, with the species most suitable to its environment the species that dominate that environment.

There are more holes in Darwin's theory of evolution than there are interpretations of that theory, but its basic precept is that one species somehow becomes another species, retaining some of the characteristics of the prior species to be used in conjunction with other chance characteristics accidentally created and incorporated with the existing characteristics because the new configuration was found to be viable in the environment.

A dog changing into a horse or an ape changing into a man (there were no dinosaurs in Darwin's world) present formidable mental obstacles. There are certainly similarities, and today with our computer software, we can morph one into another with ridiculous ease, but to attempt on a mental level the physical processes involved in converting bones and organs, hair and teeth from one to the other is just too much to endure.

When something isn't endurable, the best thing to do is find something else that is endurable. Instead of concentrating on whether such a transformation could occur, it is easier to just assume it did and focus on something else, something like the time it took to occur.

After all, we are dealing with reality aren't we, there are dogs and horses, apes and men, we aren't making up anything here, just dealing with how the facts got the way they are. Why bother trying to paint a moving picture of the transformation when we can just make up endless time, barely conceivable itself, but more conceivable than the transformation, and then say that the transformation occurred in endless time because endless time accommodates all chances?

Today, evolution from one species to the next is taken as proven fact. When anyone questions the existence of evolution, they are charged with being modern-day flatearthers, Creationists, and then the argument is couched in terms of bare religious belief or observable measurable facts. The observable measurable facts are embodied in recording the "evolution" of Darwin's finches on the Galapagos Islands during periods of drought and plentiful rainfall. Because of the type of food that survives under each condition, the environment either favors long-beaked or short-beaked finches, starving out the unfavored finches in the process.

But simply observing that one finch thrives on one set of environmental conditions while another survives during the opposite set of environmental conditions does not say anything about dinosaurs evolving into turkeys.

Why not consider evolution a process in which characteristics were evolved and encoded and thus available to all subsequent life forms to express as needed so that new species evolve by strategies which utilize combinations of superseded, existing and new characteristics as the environment dictates?

LIFE AS A CHANCE OCCURRENCE

With the discovery that boundary elements, elements at the upper range of existence, produce radiation as they break down in the given environment, and the discovery that radiation could, like the sun's rays on exposed skin, produce changes in living cells, the concept grew up that novel characteristics are the result of chance mutations resulting from exposure to radiation.

The notion that evolutionary mutation results from exposure to radiation conforms to the Darwinian need for happenstance, but why not take the simple position that life comes into existence where conditions are suitable for its existence and that characteristics do the same?

After all, life forms all around us. Just forget to put something in the refrigerator and it grows all sorts of life.

Why is there a need for chance evolution when evidence abounds that life rapidly adapts to the conditions in the environment, the short-beak, long-beak observation used erroneously to claim proof for species evolution?

Because the spontaneous generation of life smacks of biblical creationism and therefore its proposition can never be empirically supported!

As a result, life comes from spores. In fact, all life that does not come from eggs or seeds comes from spores.

What are spores?

Spores are what all life comes from that doesn't come from seeds or eggs!

A century ago, attempts were made to prove that matter would give rise to life by simple exposure to air. Sterilized matter was placed in a vacuum and observed to be dead when no life grew. When air was introduced, however, life grew in the form of mold.

Pasteur, an empirically subjective experimenter, didn't believe that life could form as a result of the exposure of lifeless matter to air and used his authority to label any experiment that was designed to demonstrate life arising from air alone the result of incompetence or fraud, the scientific terms used by Empirical

Science to control the acceptability of experimental results that disagree with desired outcomes.

For instance, hay could be boiled lifeless, placed in a sealed vacuum container, and no mold would grow. Introduce purified air, however, and mold would grow.

Pasteur claimed the purified air wasn't purified at all, but contained undetectable spores of the mold.

Because mold grew with the introduction of air, the correct procedure would have been to have the person that claimed that life couldn't arise from the interaction of dead matter and air, Pasteur, prove it by placing the matter giving rise to life in air under circumstances that no life would result.

How we phrase our questions, and the evidence we demand as answers, fashions our ignorance!

It was not possible to introduce air into a lifeless flask without producing life a century ago and it is not possible to do so today.

But because it had been subjectively decided that life could not arise spontaneously as a result of the electrical interaction between the surface atoms of the matter and the atoms that made up the air (indeed, the concept of the atom had not been thought of when this conclusion was reached), it was only necessary to make the claim that invisible, undetectable spores existed in any air that could possibly be pumped into the environment containing the lifeless matter.

When proof of the spore's existence was demanded, the fact that air brought life to lifeless matter was used as proof for the existence of the spores in the air!

Today, the claim is made that spores exist under all conditions, and that this constitutes proof that life does not arise currently, but had to arise at some point in the past purely by accident. Any application of any air to lifeless matter will result in life because all air carries spores. This has given rise to the bird poop and wafting winds biology of Modern Science and means that if life didn't develop on this planet, it developed somewhere else the way it would have developed on this planet and was empirically transported here by the comets that we see as snowballs, so it's acceptable to conclude that life developed on Earth by accident.

And it also means that the only way that a characteristic can develop is by the chance modification of an existing characteristic.

WHY CHOSE THE ACCIDENTAL PRODUCTION OF LIFE OVER DEMONSTRABLE SPONTANEOUS GENERATION?

The two experiments that show the spontaneous generation of life, Pouchet's experiment showing air bringing life to lifeless matter and the Miller-Urey sparking of chemically rich clouds with electrical discharges, are both scientific in the empirical sense and repeatable.

Why not just accept them and shorten the time needed for evolution?

After all, if matter forms around electrical flows or as a result of differentials in electrical potentials, then animate matter could easily produce characteristics to deal with new environmental situations and encode those characteristics into genes which could then become a part of the universal gene pool of life on a particular planet.

The conceptual difference is not that great, a gene that came into existence God knows how (the Darwinian point of view as opposed to the Creationist's point of view, a gene God brought into existence) compared with a characteristic, the ability to organize atoms into shells, or antennae, or eyes or paddles, that is developed in response to the need to survive within the environment, and which is encoded in a gene that is passed on to all subsequent life.

Under the former, there is no explanation for the source of life, no clear picture how one species could evolve into the next, and the need for vast amounts of time for what cannot be visualized to occur. Under the latter there is a clear explanation for the source of life, a clear picture of how species can evolve from one to another, they don't, new species just chose from the panoply of characteristics available in the gene pool to put together an organism that matches the environment, and instead of inconceivable time periods, time is reduced to the time it takes for a frankfurter to become a stick of mold.

What committed the answer to the endless eons?

THE REAL BATTLE FOR THE SOUL OF EVOLUTION WASN'T AGAINST THE FLATEARTHERS, THE RELIGIOUS OPPOSITION, IT WAS AGAINST THE PRACTICAL MEN OF SCIENCE, PRAGMATIC NATURALISTS

There was more than sufficient evidence around the globe that something on the nature of a flood had occurred and destroyed much of the life that had gone before, wiping out prior civilizations for which only remnants remained. Interpreting the evidence from such a violent viewpoint, however, would eliminate the ability to determine the history of the Earth by the fossils found in sedimentary layers, and in turn determine the age of the sedimentary layers by the fossils embedded in them, the endless chicken and egg dance that is the science of stratigraphy.

But there was a bigger obstacle to overcome than simply burying the islands of dead carcasses and tropical plants found embedded in arctic ice sheets, a greater delusion to attempt than merely ignoring the Pacific ports located fathoms beneath the sea and the petrified forests buried beneath mountains of sediment and a bolder challenge than making up idiot explanations for the vast beds of jumbled vegetation fossilized into coal fields.

The Earth was sitting in Newtonian dictated empty space, empty space was cold, laws dealing with thermodynamics required that heat flow from hot to cold, and thus, with the Earth being of a certain temperature, a reasonable estimate as to age could be made by computing how hot it possibly could have been and how long it would have taken to cool off to the point that it currently was.

Remember that evolution, like sociology today, was a nonmeasurable discipline attempting to gain influence in a world of theorists who happily deluded themselves and everyone else that measurements could prove beliefs to be fact.

As early as the eighteenth century, Buffon heated iron balls and then observed the time that it took for them to cool. Using simple comparisons of the amount of matter per volume, the uniform distribution of matter by size that Newton used in his failed attempt to prove that the attractive force was proportional to, and therefore a property of matter, he came to the conclusion that the Earth could not be older than a million years.

Something that is heated above the environment cools at a rate that is proportional to its temperature differential. A white hot ingot will cool rapidly as it emerges from the smelter, but as it cools, its rate of cooling slows. It will stay warm longer than it took for it to cool from its hottest temperature to become warm.

The chart the temperature drop produces is similar to the chart computing ten half-lives produces, the half-lives being the time it takes for half the atoms of a radioactive element to decay into an isotope. The first half-life changes half of all the atoms so the drop for the first period is sharp. The second half-life only changes one quarter of the original atoms, so the rate of drop from element to isotope starts to flatten out, with the next one eighth flattening the rate curve still further. By the time the tenth half-life is reached, the number of radioactive atoms left is approaching one percent, and the decay process, like the cooling of the warm ingot, is fairly steady.

Of course, the half-life of a radioactive element is a concept created by assuming that each radioactive emission results in an isotope conversion, an isotope itself being a concept, that the conversions are random, and that the laws of statistics can be made to apply to atoms, the atom itself being but a concept.

Buffon predated Darwin, but in the middle of the eighteenth century, the concept of evolution, and its concomitant time commitment, had already been well established. When Lord Kelvin approached the subject after Darwin had become controversial, his estimate of how long it would take for the Earth to cool to its present state was considerably lengthened, coming out at a hundred million years.

Millions of years, however, even a hundred of them, are something that people can comprehend, and when dealing with facts which cannot be known, it doesn't further the cause of obfuscation to deal with something that is even remotely understandable. When we pretend we know something, it is best that what we know is absolutely beyond the range of comprehension, and thus billions is a much better time frame to bandy about than millions. After all, the common is invisible while what is obscure, beyond the range of comprehension, inscrutable, invites interest and attention. This strategy is found throughout Empirical Science when the unmeasurable is claimed measured, billions of stars in billions of galaxies which are billions of miles away which were formed billions of years ago. The only place Empirical Science scores a direct hit with billions is in describing how many conceptual atoms make up a speck of matter, but then it goes on to defy reason by claiming to know exactly how many protons and nonexistent neutrons are in the center of the conceptual atom because it has to know what it cannot know, the rate at which the nonexistent neutrons have departed to produce the conceptual isotope used to measure the half-life of elements.

If Lord Kelvin's estimate was accepted, if he was correct, and in an empirical world it was hard to say he wasn't, then the Darwinian goose was just a goose and not a goose with dinosaur ancestors.

The argument therefore crystallized around saving Darwinian species evolution against physical proof that the Earth wasn't old enough for the endless time needed for species evolution.

RUTHERFORD'S RADIOACTIVITY SAVES THE DARWINIAN DAY

It is difficult to discern the hasty assumptions that, because of the rapid discoveries surrounding the behavior of the boundary elements such as uranium, quickly became fact and as fact dictated our understanding of the nature of matter. The watershed assumption was the creation of the neutron out of the misapplication of Newton's Third Law of Motion to subatomic particles and its subsequent use to obscure comprehension of the nature of the atom. The whole series of discoveries was topped off by the production and use of the misnamed atomic and hydrogen bombs, triumphs of the effectiveness of trial and error exploration into the workability of physical reality but quite unrelated to the concepts that went into their production.

The big bang produced by these compression devices has been universally used as proof that the silly concepts about radioactivity created on the fly while facts were being rapidly uncovered are basic laws of the universe never to be tampered with by the mind of man that made them up.

The basic principle of radioactivity is that the more active the "radio", the shorter the half-life of the element producing the radioactivity. "Radio" is short for radiant energy, and radiant energy is any energy that radiates out in rays. Heat is radiant energy, and heat is measurable by putting your hand in front of a fire. But heat is an imponderable. According to Modern Science, heat doesn't exist between its source and its effect, so it can be distinguished from the type of radiant energy produced by radioactive elements, or elements that have been made active, activated by some sort of bombardment. Radioactive elements produce measurable rays.

For purposes of this particular panel of the evolutionary cartoon, however, radioactivity produces the imponderable, heat.

Now, if the Earth had radioactive elements down there somewhere beneath the surface producing heat, then Lord Kelvin's clock would never run out because the heat that would be lost to space was not being lost to space, it was being replenished by the heat produced by the radioactivity of the elements inside the Earth.

This sort of hazy thinking, that the Earth is not losing heat to space because it is being kept warm by its internal heat which is produced by radioactive elements beneath its surface mirrors the understanding of heat, which is an imponderable and doesn't exist between its source and its effect, there being no effect of the Earth's heat in space, and of radiation, which is heat made ponderable by clicks in detectors such as Geiger Counters.

THE BILLIONS OF YEARS IS IN THE CLICKS

When matter was found to emit something other than heat, the emissions were measured as scintillations, little sparks of light, which, strangely enough Modern Science also considers an imponderable which doesn't exist between its source and its effect. These scintillations were produced when the radioactive rays or particles came in contact with a detector, usually zinc sulfide painted on a glass slide.

Much of our measurements involving subatomic size and amount, and thus our knowledge of the nature of subatomic particles, is the result of people sitting in darkened rooms counting these little scintillations and noting their brightness, thus marking the subjective nature of Empirical Science.

Geiger found a way to electronically monitor the scintillations, creating a circuit the rays or particles produced which was connected to a noise maker, the click, click, click of a Geiger counter, and which can, being electronic, be monitored by all of the marvelous computer driven devices we produce today.

Fermi found that nonradioactive elements could be activated. Bathing nonemitting aluminum in beryllium emissions would result in radioactive aluminum. The activation, however, was only temporary.

Not surprisingly, different elements were found to have different lifetimes when activated, giving rise to Nobel attempts to integrate radioactivity into the Periodic Table which organizes elements by weight and electrical property. To accomplish this, artificially activated elements needed some sort of measuring rod to determine their ability to become radioactive. The 2.3

activated life of silver was chosen.

The life of an activated element, however, could not be used as a measuring rod because it could not be applied to elements which were found in nature to be radioactive. There was no way to measure the actual life of naturally radioactive boundary elements.

But because activation was considered to be a product of neutron capture, and neutrons had weight, neutron capture would increase an element's weight, and because radiation was considered the opposite, the loss of neutrons, and because weight was, as a result of Newtonian mistakes inextricably confused with mass, radioactivity was thought to involve the loss of mass which meant that as radioactive elements emitted their radiation, they lost mass and became isotopes.

Thus, a radioactive element was made up of atoms which were being converted to the atom's isotope. Because it was unknown which atoms were being converted to an isotope at any one time, the process could be described with the mathematics of statistics in which a certain percentage of the atoms lost their radioactivity in any given period.

The period used was the half-life of the element, the period during which one half of the atoms in the element could be expected to convert from the atom to an isotopic atom as a result of emitting radiation.

This led to the basic law of radiation, the more active the radiation, the shorter the half-life, which parallels the white hot ingot, the hotter the thing is, the more rapidly it cools off.

In what is one of the most ingenious illusions in Empirical Science, the mathematical formula devised to statistically compare the artificial activation of elements was then applied to nonactivated boundary elements to determine how long they would have been radioactive had they started out with a full activated life.

The heaviest elements in nature, those at the upper boundary of existence, survive because they are electrically balanced in combinations with other, lighter and therefore more stable, elements. They do not emit radiation. Isolate these heavier elements, however, and they evidence a mild radioactivity. Many nonradioactive elements can be made radioactive for short periods. The more radioactive an activated element is, the shorter the artificial activation and the shorter the element's half-life. Kick a rock loose from its pitchblende and find it to be mildly radioactive. Apply the measuring rod created to compare artificially activated elements and what do we get?

An extraordinarily long half-life!

The Darwinists needed eons of time to explain species evolution, but physical science showed that the Earth didn't have eons of time, it would have cooled off long before incomprehensible species evolution could have occurred. But wait, it did have eons of time because the Earth's heat was constantly being replenished by radioactive heat. And those eons could be demonstrated by taking a rock on the surface of the Earth that is mildly radioactive, that supposedly has a high proportion of isotopic atoms, and then computing backwards to determine the age of the Earth, which fortunately turned out to be the half-life of the barely radioactive boundary element found in the rock, four and a half billions of years.

This is sort of like coming across an ingot, processing it so that it becomes a billionth of a degree warmer than its environment and then using its rate of cooling to compute when it was forged.

But more silly, the very radioactive heat that the Darwinists used to keep the Earth warm for the billions of years needed for species evolution would also be the same radioactive heat that continually reactivated the boundary element used to put the figure billions to those years, destroying radiometric dating as a clock!

The radioactive Earth is great when the cartoon panel has to demonstrate why the Earth hasn't cooled off in the cold of space but has to be discarded when the cartoon has to use radioactive elements on the surface of that continually radioactively reheated Earth to prove how long the Earth has existed because radioactive heat resets the radioactive clock by reducing the proportion of isotopic atoms.

In short, if the radioactive heat used to warm the Earth were factored into the radioactive clock, the Earth's age would be back to somewhere around Buffon's million years.

When two panels of the cartoon are inconsistent, the Empirical Rule is to focus on only one panel at a time!

But then radioactivity is so useful, useful being the Empirical measure for maintaining fantasies. It's rays provide the chance mutations that drive evolution, its heat gives these chance mutations the time to produce species evolution and the isotopes it produces configure the clock that cloaks the whole process with an appearance of unassailability.

THE MORE PROOF IS LACKING, THE MORE PROOF THERE IS

The age of the Earth is unknowable. When something is unknowable, there is no amount of proof that can be generated to demonstrate it.

There is, therefore, no proof that the Earth is older than a million years, let alone billions of years old. In fact, Buffon's experiments are probably closer to the truth than Lord Kelvin's Darwinian influenced estimates which, under the glare of reality, he quickly reduced by three quarters.

There is no proof for the proposition that the Earth has been the same temperature since its inception. In fact, the proof is quite the contrary. Things in physical reality cool off to the temperature of their environment, and the environment that the Earth occupies does not have temperature, temperature being a measure of activity and space being empty. If radioactive elements kept planets warm, the moon would shine when it wasn't full.

There is no proof that the history of the Earth is governed by uniformitarianism, the lack of action other than imperceptible action over long periods of time. In fact, the evidence is quite the contrary, with Earthquakes and volcanoes changing the geography of the Earth overnight even today. Given the fact that heat drives geographical changes, *and in more ways than presently suspected*, a hotter Earth meant a more violent geographical landscape, a landscape which could see the formation of mountain ranges as rapidly as television crews see the collapse of California freeways today.

There is no proof for a worldwide sedimentary build-up that would produce a uniform dating system. In fact, the evidence is quite the contrary, with sedimentary build-up being what it is, the result of local conditions. The historical "record" of the Siberian tundra is not writ large under a backward barbecue in Queens.

There is no proof that life came into existence accidentally. In fact, evidence is quite the contrary, with Pouchet's experiment consistently showing lifeless matter giving rise to life with the introduction of common air. Because we make up spores, then associate those spores as the source of bacteria and viruses, fungi and mold, then claim that because we find viruses and bacteria, fungi and mold under all conceivable environmental circumstances, the spores that cause them, spores we have made up because we have assumed that the viruses and bacteria, fungi and mold cannot come into existence spontaneously, exist in all conceivable environmental circumstances and thus air can never be provided the dead matter without the conceptual spores that give rise to the viruses and bacteria, fungi and mold doesn't change the reality that life comes into existence daily all around (and in) us.

There is no proof that species evolve one to the next, and in fact, the evidence is quite the contrary, with new species appearing daily. Using the word "mutated" to describe elementary life forms that evolve to meet new environment conditions, the actual proof of evolution, in place of evolution because we believe that evolution only occurs over eons begs the question. Life is the organization of atoms and molecules of atoms around electrical flows, and when the environment permits electrical flows around which atoms and molecules of atoms can organize, those atoms and molecules of atoms will organize immediately, and continue to organize to take advantage of all subsequent electrical flows generated either by the environment, or by the organized atoms and molecules of atoms in the environment. Life forms rapidly and completely, produces advantageous characteristics, encodes those characteristics in genes, and, when it become possible to replicate, passes those genes forward to all subsequent life forms to use or not use as the environmental needs dictate. The more complex the life is, the more strategies it can bring to bear in overcoming environmental obstacles and the faster it will evolve new characteristics.

And the proof for radiometric dating?

Radioactive elements have to be extracted from a material such as pitchblende. Radiometric dating assumes that the elements have been active while they are in molecular balance with the elements that make up the pitchblende. Zinc has a melting point of 420C, but when it is placed in molecular balance with copper, the resulting brass has a melting point of around 1050C. Saying that a radioactive element found in molecular balance with pitchblende has been radioactive since the pitchblende was formed is similar to saying that the zinc in brass begins to melt at 420C even though the brass itself continues intact. Add this to the absurd conclusion that radioactive elements beneath the Earth's surface are continually reheating the surface of the Earth, but are not affecting the radioactive elements used to date the surface of that Earth, and you have what one would expect from a belief system that actually believes that everything in the universe is going in straight lines.

The litany of Sagan, the most famous delusionist of Empirical Science, was that extraordinary claims require extraordinary proof. Empirical Science requires that claims at least be demonstrated. When the Empirical Scientist produces the lab notes that show the half-life of Uranium-238 turned out to be demonstrated at 4.5 billion years, then that claim can be accepted as fact. Until then, it is merely a fantasy created to fill in the penultimate panel in a totally absurd mental cartoon.

Radiometric dating dates our beliefs, not the age of the Earth. And that is the most important test of its validity, because it integrates itself into a tapestry of fantasies that, taken together, become unassailable. After all, the basis of Modern Science is the probability that a conclusion supported by many interlocking and mutually reinforcing clues can be close to certainty, even if every clue by itself is less than definitive, leading Modern Science to its overriding belief that the existence of many different strands of otherwise unsupported evidence can transform a belief into a fact.

The more individual fantasies fail support, the greater the validity of the overriding belief giving rise to those fantasies!

What we want to believe is what we believe to be fact because we want to believe in an incomprehensible species evolution, the incomprehensible time it took to produce it, because we want to believe in a uniform environment stretching back into the unimaginable past and therefore into the unimaginable future, safety for ourselves and our progeny.

Above all, we want to return to our ego based religious precept of man as central to the universe.

It may be empirically measurable that we are not at the center of the universe, but we can produce a belief system that is just as good, that has us alone in the universe, an accident of chance unlikely to have ever been duplicated and unlikely to occur ever again.

After all, being alone and being unique is just as good as being at the center.

THE CONSEQUENCES OF A BILLIONS YEAR OLD EARTH

When facts are unknowable, how we phase our questions and what we demand as proof will determine what we end up knowing.

We are literally between heaven and hell, and neither appear to be desirable. If we look at the surface of Venus, we can visualize life forming in forming puddles of water as the planet cools much as it had to have once formed on the surface of the cooling Earth.

But if life comes into existence when conditions in the environment permit its existence, then the Creator has to be pushed back to the recesses of the universe, and with Him, our uniqueness, our aloneness, our centrality!

On the other hand, looking to the heavens, we see the moon, a dead cold rock. If we believe that the Earth is cooling off, that it was once hot like Venus, and perhaps even a miniature sun, then in staring at the moon we are staring the fate of the Earth, and our own fate, in the face.

Reality is difficult to perceive when viewed through our desire for something more than reality.

As a result, we phrase our questions and demand our proof as they relate to what cannot be known, and do so in the name of asking questions about objective reality that are not capable of being answered. We ask and answer questions of origin and end, of how large and how small, and in answering those questions, limit our knowledge about the questions of obvious reality, questions of what makes objects fall, what makes planets rotate and orbit, what is light and electricity and magnetism.

We assume away questions dealing with reality, claiming gravity to be a property like color, motion in space to be historical, light and heat imponderables that have no existence, electricity a moving charge and magnetism the product of molecular magnets.

By busying ourselves answering questions the answers to which are unknowable, we give ourselves the appearance of having answers when we know nothing about the matters that are absolutely essential to our survival and for which questions are answerable, movement and force.

Thinking that we are unique, that we have taken billions of years to get to where we are, we think we will be here for billions of years more so long as we don't shit in the drinking water, and we lose sight totally of the fact that the universe grinds on regardless of how we delude ourselves.

If we occupy ourselves with our own little ant hill and never concern ourselves with the possibility of surviving beyond it, we will never evolve the characteristics that will allow us to survive beyond it and will simply grow cold when it grows cold and die.

9. THE NONEXISTENT NEUTRON

THE DIVIDING LINE BETWEEN EMPIRICAL SCIENCE, IN WHICH BELIEFS WERE BASED ON MEASUREMENTS, AND MODERN SCIENCE, IN WHICH BELIEFS ARE BASED ON BELIEFS!

AN INCREDIBLE CONCLUSION

The neutron was created by default.

It is the creature of a passel of misconceptions that lay hidden at the foundation of the scientific thought processes prevailing at the beginning of the twentieth century. Experiments leading to the creation of the neutron, performed over the first third of the twentieth century, exposed these misconceptions to the light of day. However, rather than examining them for what they were, bald assertions of belief, the experimenters blinked and, in the resulting blindness, created the nonexistent neutron.

Had the times been normal, the nonexistent neutron might well have remained nonexistent, or a least theoretical. However, with fast breaking technical developments in physics rapidly sweeping theoretical musings about the nature of matter into the looming World War II effort, the misconceptions quickly sank back to unexamined core beliefs. With no other explanation for the energy that was being discovered in boundary elements related to uranium and with the vast efforts poured into producing technology that could utilize this potential energy to further the war effort, the neutron was mindlessly accorded the status of fact, being the litany used to avoid explaining nuclear energy much as the words "gravity is a property of mass" are used to avoid explaining the cause of falling objects and the words "angular momentum is conserved in closed systems" are used to avoid explaining the cause of the Earth's rotation.

One of the misconceptions that led to the belief in the nonexistent neutron is the law dealing with the conservation of energy which states ideally that in an isolated system the amount of energy remains constant but which is generally taken for the proposition that energy can neither be created nor destroyed. The law lends itself to the construction of mathematical models used to explain how systems behave under a given set of circumstances but has little to do with reality because the disciplines that wield it have little knowledge of, or interest in understanding the nature of energy. When new energy sources, such as electricity or nuclear energy are uncovered, they are calibrated to existing energy sources such as coal or oil by the amount of work they can accomplish or the heat they are estimated to produce and an explanation for their source is glossed over with meaningless words such as "moving charges" and "polarity" or "the equivalency of matter and energy." Heat itself is reduced to a form of energy involving motion, and while thought to be transported through empty space by radiation, is considered to be an imponderable while so moving, which is to say it doesn't exist between its source and its effect.

The conclusion that heat is an imponderable, that light is also an imponderable, and that the two are separate and distinct phenomena is the second misconception involved in the production of the neutron out of whole cloth. If heat is thought of as nothing more than motion in the matter that is producing or experiencing it, then it is not considered to be something in and of itself. It is not considered to be the movement of something from one place to another, but merely the transportation of an effect through space by undelineated radiation. Heat and light radiation don't exist between their source and effect and therefore become invisible in analyzing what might be at the basis of measurable physical effects.

A more subtle stupidity, Newton's Third Law of Motion, that for every action there is an equal and opposite reaction, is pretty obvious on a physical basis. If it weren't true, pool halls would go out of business and the rockets we shoot into space would never leave the ground. The law, however, describes the behavior of physical matter that is restrained by the physical force of attraction generated at the Earth's surface. The mechanics of this force, labeled gravity, and thought to be a property of mass, remain elusive, and as such, the way that gravity affects matter so as to restrain its motion is unknown. Any assumption dealing with how it acts on an atom, then, is premature. The conclusion that gravity restrains the electrons that orbit the nucleus of an atom in the same manner that it restrains the units that make up the nucleus is unwarranted.

A misconception passing the range of conscious analysis is to then go on and claim that Newton's Third Law of Motion applies to the movement of the electron itself. Without an understanding of the nature of energy, the motion of the orbiting electron remains unexplained. The same lack of mental acuity that allows acceptance of the statement that the force that moves released objects at an accelerating rate toward the surface of the Earth is a passive property of the matter doing the moving also results in the failure to discern that *motion itself might well be an appropriate property, while causing motion is not.* If in fact the motion of the electron in orbit around an atom's nucleus were the result of the electron's property to move at, say, the speed of light, then there would be no question that Newton's Third Law of Motion failed applicability. The electron itself would not be subject to the pull of gravity and its motion would be with respect to the nucleus of the structure of which it was a part, and not a particular planet toward which that nucleus was accelerating.

And what is the motion of the orbiting electron with respect to? What force would cause an electron's straight-line motion, with Newton's First Law of Motion here applicable if nowhere else, to bend into the circular motion that defines its orbit around the nucleus?

Polarity, a misconception of application, is probably the most unexamined concept ever carried thoughtlessly into unrelated domains of knowledge. This concept, dealing with the physical behavior of magnets has crept through hazy explanations for static and current electricity into the concepts used to explain the structure of the atom. The general rule, that likes repel and opposites attract, created as an observation to describe the physical behavior of magnets and codified as mindless law, was an insurmountable obstacle to figuring out what was happening in the subatomic realm where it has no applicability but where, because it lay hidden in the thought processes, was accorded supreme authority. If likes don't attract in the subatomic world, then all sorts of particles have to be created to hold structures together that would naturally form together if likes did attract. For instance, if there is an orbiting electron, and the electron is negative, the statement mindlessly carried over from watching toy dogs kiss requires that there be an oppositely charged particle created for the electron to orbit.

The final misconception underlying the creation of the neutron is the misperception that because we produce scales that allow us to measure the relative effect of the attractive force on matter, and then label the effect of that force on matter as weight, that the matter itself has weight. This misconception had its origin in Newton's corpuscles of matter. In order to prove that gravity was proportional to and therefore a property of matter,

Newton had to mathematically compute the amount of force, and the only way that he could do this was to compute the amount of matter producing the force. To compute the amount of matter producing the force, he had to assume that matter was distributed uniformly by volume throughout the matter producing the force, the Earth and the moon. The exercise failed proof, but later day gurus saved Newton's matter/gravity hypothesis by decreeing that if volume couldn't predict motion, then motion could be used to determine the density of volume, the matter's mass. This resulted in the notion that matter has intrinsic weight when in fact the only thing that gives matter weight is the force that is causing it to move. The weight is the measurement of the force that it takes to overcome the force that is causing the matter to move. It is merely the amount of surface force, calibrated by spring, it takes to overcome the force of the Earth's attraction. Matter itself is passive in this struggle, just falling in the force if it can, and coming to rest against surfaces that are sufficiently strong to overcome the force that is making it fall.

By making the mistake of thinking that weight is inherent in matter, that it is a property of the particles that make up the matter, it became necessary to make up particles in matter to account for its weight when all that's necessary is to have a particle in the nucleus upon which the force of attraction can act.

A whole panoply of misconceptions, then, have gone into the conclusion that the conceptual neutron, a particle made up simply because there appeared to be no other explanation for what was measured to be happening, existed. There was no other explanation for what appeared to be happening because a commitment had been made to a belief that had no basis in reality, that the conservation of energy applied in a situation where energy itself might be a property of the particles being measured. Because the energy was not viewed as a property of the particle and therefore a result of the particle being released, the effect of the released energy was viewed as a reaction. Using Newton's Third Law of Motion to justify observed effects that required the creation of an unmeasured energy because otherwise the energy would be viewed as created, the neutron was created out of nothing to explain the source of the energy, the action that had to be producing what was viewed as a reaction rather than a release.

Once the neutron was created to explain a source of energy *where no source of energy was necessary*, the failure to understand the simple process involved in the production of energy created additional complexity because of commitments made on the nature of motion, that motion requires energy rather than being energy, on the nature of heat and light, that heat and light don't exist between their source and effect, on the applicability of polarity to the structure of matter, that matter is held together by particles that repel each other and on the nature of weight, that weight is intrinsic in matter rather than the measurable effect of gravity on matter.

With such well rooted concepts, invisible to the intellectual process, operating, -- "if you don't believe in the neutron, go stand under a hydrogen bomb" -- attempting to explain the creation of the nonexistent neutron when its existence is taken as fact should be as difficult as clarifying a caballa to Casper the Ghost.

However, it's not as difficult as one might imagine.

IMAGINE REALITY!

Imagine holding an unlit match about two feet away from a lit match.

The heat and light coming out of the lit match are expanding in an expanding sphere that is moving out and away from the surface of the match, the match itself being consumed as the heat and light move out and away from it. Of course, what is making up the heat and light cannot be the same thing that is making up the match because it has been empirically proven that heat and light do not have mass and are therefore imponderables with no existence between their source and effect. However, if a property of heat and light is motion, the definition of energy, then they don't need mass to exist.

Because heat and light expand over the surface of an expanding sphere after they leave the lit match, they diminish inversely with the square of the distance over which they expand, the distance from the surface of the match. By the time the heat and light get to the other match, two feet away, there is not enough heat and light left to produce an effect on the second match. Some of the heat and light simply bounces off the surface of the match, some of it penetrates the surface a small distance, and some actually goes right on through the unlit match to join the heat and light that has passed around the match, but the unlit match remains unlit because of the weakness of the expanding sphere of heat and light, which has diminished too much to ignite the match.

Now imagine moving the unlit match toward the lit match.

More of the heat and light from the lit match penetrates the surface of the unlit match, and more of the heat and light passes into and through the unlit match.

One of the things that we know, or should know about the unlit match is that it is made up of atoms which have clouds of electrons orbiting them. The electrons in these clouds of orbiting electrons are held in their orbits by an electrical charge in the nucleus of the atom.

If instead of viewing the heat and light as imponderables, we view them as flows of electrons, which, after all, are conceptual particles made up to describe an effect, then what might the effect of these flows of electrons be as they pass through the unlit match?

We know one thing about flows because we can sit by the side of a river and watch the water flow. We know that at every point in the flow of water, there is water. It is not the same water, but there is always the presence of water in a flow of water.

So, if we think of the expanding sphere of heat and light as made up of flows of electrons, we know that at any point in the flow, there will always be electrons.

If we make the simple assumption that electrons are orbiting the nuclei of the atoms that make up the match because one of the electron's properties is the property of motion, then we know that Newton's First Law of Motion would cause an orbiting electron to take off if the force that was holding it in place, restraining its straight-line motion, were to be removed.

We know that we can't remove a force that exists. However, if we can't remove something, how about replacing it?

The electrons in the expanding sphere of heat and light have a current presence because at any point in a flow there is always an electron. If a flow of these electrons were brought next to the nucleus containing the orbiting electrons, then the force in the nucleus that was being used up attracting the orbiting electrons would be occupied by the flow passing by and with nothing left to hold the orbiting electrons in place, they would depart in a straight line.

This is what happens when something burns, flows of electrons are replacing orbiting electrons which then join the flows, producing a conflagration in matter.

Another way of looking at the process is to say that the orbiting electron has more energy than the electron in the flow because the energy of the electron in the flow is being used up moving the flow. As the flow is stationary with respect to the nucleus, the orbiting electron is no match for the electrons that make up the flow. The orbiting electron is already in motion so, no longer having an exposed charge in the nucleus to which it can be attracted, it regains its full motion and departs the nucleus in a straight line. (One of the main reasons Modern Science is such a morass of inconsistent concepts is that it requires force in the subatomic world where motion can be accounted for by assuming it to be a property of the particle that makes up matter, and doesn't require force in the measurable world, where the predominate force, gravity, is assumed to be a property of the matter toward which objects move and the movement of matter in space requires no other force, because the matter is assumed to be moving in a straight line but for the force of gravity. These are the fervent beliefs of Empirical Science even though, in reality, a match produces more energy than it takes to ignite it and objects in space are neither moving toward one another nor in straight lines.)

Viewing combustion as a process that releases the energy stored in matter by replacing one field with another, as the unlit match moves closer to the lit match, the flows of electrons that make up the heat and light of the expanding sphere of the lit match replace the electrons orbiting the nuclei of the atoms of the unlit match.

As the unlit match moves deeper into the field of the lit match, this replacement increases in strength with the square of the distance until the rate at which the electrons are departing the nuclei of the atoms of the unlit match causes the match itself to ignite and itself produce an expanding sphere of heat and light.

This is field replacement, it is produced by eliminating the application of unfounded assumptions concerning motion, mass and polarity to the subatomic world and it is an alternate explanation to the neutron to explain the source of the measurable energy that the neutron was created to explain.

Instead of viewing radiation as a process of releasing energy, the normal conclusion that would be made by simply looking at what happens in physical reality, the effect of radiation, the release of energy, was viewed through the lens of the action and reaction of matter, intrinsic weight, the conservation of energy, the belief that likes must repel and thus the belief that the subatomic world behaved like loaded balls on a billiard table.

When a reaction occurred with no apparent energy source, instead of examining the assumptions about momentum underlying the nature of radiation and in the process, adopting a hypothesis other than Newton's Third Law of Motion, an invisible cue ball, the neutron, was created to explain the unexplained energy. The neutron was accepted as a substitute for the unexplained reaction because a particle could bring energy to the transaction. While there was no explanation for the source of the energy the neutron brought to the transaction, the source of the neutron's momentum, the neutron's energy didn't need explaining because momentum, the cause of straight-line motion, had been assumed away from the time of Newton, who assumed it away because he had no explanation for planetary movement.

By assuming away the source of the neutron's energy by saying it is derived from its momentum, obvious physical phenomena such as combustion go unexplained and concepts instead are entombed in words to prevent acute questions from bothering muddled minds.

WHAT WE KNOW AND DON'T KNOW ABOUT MATTER

We have two ways to analyze matter and both of those ways give us insight into the structure of matter. One way is to put something on a scale and weigh it. The measurement of elements by weight with a scale produces a list of the elements going from the lightest to the heaviest.

If we then measure the electrical potentials of the elements that make up the list, we find that they produce repeating patterns, the repetition giving rise to the periodic placement of those elements by the varying strength of their charge so that the elements form a table that can be used to predict the existence of other elements.

Both of these properties of matter, weight and electrical potential, are obvious in physical reality.

That matter is held together by a force is apparent and that different matter requires different levels of force to move it against the attractive force produced by the Earth is also apparent. Thus, the periodic table can be explained simply by viewing matter as being made up of atoms with nuclei composed of a unit on which the attractive force acts and which itself produces some sort of attractive force that holds the units, and the atoms together.

We know that the attractive force that causes matter to move is not the same attractive force that holds the matter together because, instead of a steady progression from light to heavy, the force that holds the matter together varies in repetitive cycles as the elements become heavier.

What we have made out of this limited knowledge is quite remarkable!

Atoms are composed of a nucleus that contains protons and neutrons. Because atoms are conceptual, so small that we only imagine to be able to image them by using the variations in electron beams bouncing off the surface of matter to form a picture of the variations that occur between that surface and the environment it occupies, we think of specks of matter as made up of billions of atoms.

But when it comes to producing a picture of the atom itself, we have no hesitancy in coming up with precise figures with respect to the various conceptual particles we make up to make the atom up.

Thus, the lightest element, hydrogen, has one proton and one electron orbiting the proton with the next lightest element, helium, having two protons and two neutrons in the nucleus and two electrons orbiting the nucleus. As the nucleus becomes larger, the number of protons, neutrons and electrons becomes larger, with the number of negatively charged electrons, which produce the repetitive periodicity of the periodic table, matched by an equal number of positive protons in the nucleus, and neutrons added to account, along with the protons, for weight. As the weight of the element increases, however, the electrical potential that produces the repetitive periodicity of the elements dictates that there cannot be more than eight electrons in the outer shell. The electrical potential, or valence, is an atom's ability to combine with other atoms. The numbering system that represents electrical potential is represented by assigning a specific number to the atom, with "1" being highly reactive and "8" being nonreactive. The logic for this system is that atoms with a valence of less than "8" can combine with other atoms until the atoms of the resulting molecule have eight electrons orbiting them in some combination of orbits. This state of matter, in which no charge is measurable, is inert, electrically neutral. With the charge of the electrons balancing the charge of the nucleus, the external charge of the matter is neutralized. If the numbering system is converted to a comparative charge, then the lower the valence, the higher the charge.

Hydrogen nuclei therefore carry a higher charge than helium nuclei, helium nuclei forming into an inert gas.

As the number of electrons and protons increase with weight, the electrons are layered into shells, with the outer shell containing from one to eight electrons, and inner shells added as necessary to accommodate the increasing numbers of electrons.

The number of protons determines the element, with neutrons, controlling atomic weight, capable of producing isotopes of an element as they are gained or lost.

Because the further we move from reality when we attempt to describe reality, the more exact we can become, describing what occurs in the nucleus is quite precise. When an element undergoes what is called alpha decay, it emits an alpha particle which is a bundle consisting of two protons and two neutrons. This is the nucleus of a helium atom. The remaining nucleus now has two fewer protons, and thus two too many electrons in its outer shell. These two electrons sort of wander off. The result is not an isotope, but a different element, one lower down on the periodic table because of the change in weight and electrical potential. Thus, atoms of uranium-238 emit a helium nucleus and become atoms of thorium-234. An atom can also lose or gain an electron and in the process, gain a charge, become ionized.

This numbering system is not even considered representative, with say a uranium atom actually having 238 zillion protons and neutrons and the thorium atom actually having 234 zillion, with the atom capable of emitting four zillion protons and neutrons and approximately half that number of electrons. Once the periodic table was constructed and numbers assigned to it, all matter, predicted or miscategorized, had to subsequently be fit into the numbers on the table. Thus, when we think of radiation, we don't think of a lot of atoms each emitting a lot of particles, we think of a lot of atoms each emitting a single particle or maybe four. This atomic preciseness is necessary in order to produce radiometric dating which, based on the conversion of atoms to isotopes, is needed to justify the time thought necessary to produce the species that is making up the specific contents of the atom.

Beta rays, as opposed to the alpha decay that produces a helium nuclei, result when one of the neutrons in the nucleus emits an electron, converting itself in the process into a proton. This produces a change in the electrical potential of the atom without changing the atom's weight. When an element produces beta rays, it's producing a flow of electrons, but because a flow of electrons is a lot of electrons, and the protons in an atom, limited in number, are incapable of giving up a lot of electrons, these electrons have to come from different atoms with a whole bunch of atoms having to contribute to a single beta ray.

How this might occur is anybody's guess.

Finally, gamma decay involves the protons and neutrons in the nucleus engaging in a little hoe-down, emitting electromagnetic radiation as they dosi-do to a Maxwellian two-step. Because neither the weight nor the electrical potential of the atom is affected, this electromagnetic radiation, in the form of gamma rays, is, like heat and light, an imponderable, without existence (unless it happens to be bullying electrons around in the Compton Effect.)

In reality, the atom is probably made up of electrons which, unaware of our Nobel laws dictating their existence, and thus not required to repel each other, form into units of matter which, when the last electron that can be attracted is attracted, form into nuclei which, when the last unit that can be attracted is attracted, attract either an orbiting cloud of electrons, other nuclei, or both to form into matter.

The proton is merely a neutron that has lost an electron, just like our current concept, except that the proton contains billions of units made up of electrons and there is no neutron. A hydrogen atom is a proton with an orbiting electron, just like our current concept, except the proton is actually a conglomeration of billions of units made up of electrons and the electron is a cloud of trillions of orbiting electrons.

The electrical potential determines the size of the orbital clouds and in turn is determined by the charge produced by the units in the nucleus, a charge that, because it depends on the remaining charge after the units in the nuclei come together, gains and wanes as the nucleus becomes larger, and the number of units in the nucleus directly determines the weight. Thus, while helium is four times heavier than hydrogen, the orbital cloud of a hydrogen atom contains twice the orbital electrons of the helium atom because, in accordance with the valence numbering system, the charge of the hydrogen nucleus is twice the charge of the helium nucleus.

If there are no neutrons in the nucleus, then exactly where did the neutron come from?

FROM LIGHT TO SPARKS TO DARKNESS

Thomas Edison is credited with inventing the light bulb. The light bulb is the production of resistance to a flow of electricity in a vacuum and the resulting release of the resisted electrons. A battery or generator produces an electrical flow in a circuit. A light bulb consists of the vacuum contained in a bulb and placed as a load on the circuit. A resistance in the bulb, the filament, retards the flow of electricity. When the electrons that make up the flow of electricity in a circuit encounter the resistance of the filament, they are organized into heat and light, and in the process are used up so that they have to be replaced in the circuit, giving the electric company a basis for sending us our monthly bills.

For a decade or so after the development of the light bulb, this was exactly what was thought to be happening. Led by the German physics community, light was considered to be made up of the same thing as the electricity that produced it, with the mechanical process of the filament merely being one of organizing the electricity into light. Unfortunately, the British Royal Society had already decided some years before the discovery of the light bulb that light didn't exist, it was merely a wave, a disturbance of some sort of undetectable medium. Light could not therefore be coming out of the filament. The light might be coming out of the filament in the same way that the heat, a phenomena different from light, was coming out of the filament, as a form of electromagnetic radiation but neither heat nor light were ponderables, had mass, weight, existed, so that the light, while caused by the electricity moving through the filament, was unrelated to the electricity itself.

In order to empirically prove that light was not the same as the electricity that produced it, and settle the argument in the process, the filament was removed from the light bulb, leaving a gap in the circuit within the vacuum. The point at which the electricity entered the bulb was named the cathode and tagged as negative while the point at which the electricity would have left the bulb was styled the anode, with its analog being positive. If the anode was fashioned into a cylinder, then the electricity from the cathode could be directed down the center of the cylinder where it would exit the end of the bulb, made into a tube to accommodate the shape of the cylindrical anode.

Whatever was coming out of the tube, however, was invisible, and only by pointing the tube at something that would evanesce could it be determined that the cathode tube was producing cathode rays.

Using a glass pane covered with zinc sulfide as a target that would evanesce, the rays emitted from the cathode ray tube could be observed in the form of, surprise, light.

So we got a light bulb that produces light, we modify it so it doesn't produce light but something else, something we can't see, to prove that the light isn't the electricity that is producing it. We then make what we can't see visible, we get it to produce light.

And we use this to prove that light isn't made up of the electricity that produces it!

Light was considered to be different from what was causing the spark because the location of the spark could be moved by moving a magnet next to the cathode ray tube.

Empirical Science has no idea what the structure of light is, or why, because light diminishes as it expands, and does so uniformly, the structure of light might involve, as does the structure of matter itself, the neutralization, the balancing of the charge produced by the particle that makes it up. However, being ignorant of facts, which are eliminated willy-nilly in the name of modeling closed systems, has never stopped Empirical Science from establishing mindless conclusions that forevermore rule its thoughts, blinding it to the light of day.

Finding that cathode rays could be deflected by a magnet was decisive in dictating the conclusion that the flow coming out of the cathode ray tube was not the same as light.

Thomson assumed that the deflection meant the rays were made up of negatively charged particles and borrowed Stoney's name for a charged particle, the electron, as their name so that cathode rays *were* made up of the electricity that produced them.

Light, however, was not a particle, it was an imponderable. As it could not be deflected by a magnet, it was chargeless.

Light was therefore not made up of electrons and was not the same as the electricity that produced it!

WHY WERE PARTICLES NECESSARY?

The sparks produced by flows of electrons being emitted by a cathode ray tube were, in the spirit of building a body of knowledge that could simulate understanding by being words, named scintillations. Although rays from uranium were discovered by their ability to penetrate photographic plates, they could also produce scintillations on zinc sulfide coated panes of glass. By placing "absorbing" metal foils between the uranium and the zinc sulfide coated glass, the radiation's energy could be determined by the number of foils the radiation could pass through and still produce a spark on the zinc sulfide coated glass. The more energy the radiation had, the more energetic the particle that made it up was considered to be.

This sort of analysis assumed momentum in the radiation, so if motion is going to be assumed, it might as well be assumed for one particle, with the different effects being the result of the way the particle is configured or the way it is released. Empirical Science, however, can't measure the cause of motion, only the motion, so the motion exists and the cause is ignored. This allows it to create particles that assume momentum to explain the missing force when applying Newton's Third Law of Motion to the subatomic world and ignore reality, the field emitted by combusting or otherwise disintegrating matter, as a source of force.

The three forms of radiation, alpha, beta and gamma, were analyzed by assuming their energy was proportional to the number of metal foils, or in the case of gamma radiation, the thickness of a lead sheet, that the radiation could traverse with charge intact, the charge being what sparked the coated pane. The first, alpha radiation, was considered to be low energy radiation because it was quickly absorbed. It couldn't travel through many layers of foil and still spark the coated glass. It was thought to be composed of alpha particles, helium atoms without their orbital electrons. Beta radiation was more energetic, passing through more layers of absorbing foil and was analogized to the cathode rays because the two survived a similar number of absorbing foils while producing similar sparks on the coated glass. The beta particle that produced beta radiation was labeled an electron and the terms used interchangeably. The third type of radiation, gamma radiation, was very energetic, the most energetic of the three and thus analogized to Maxwell's electromagnetic radiation, placed, because of its penetrating ability, at the short wave, high frequency end of the radiation spectrum.

Because gamma radiation was considered to be a part of the electromagnetic spectrum along with heat and light, and therefore an imponderable, there was no gamma particle.

Why were some types of radiation considered to be made up of particles and others not?

During the nineteenth century, light was considered to be a wave and a wave required a medium for its existence. When the Michelson/Morley experiment showed the medium to be a figment of the scientific imagination, Maxwell's dummied up proportionality equations were adopted to save the concept of light as a wave. These equations produced a fuzzy idea that attempts to explain light's motion. Because inductances were similar to magnetic lines of force -- breaking either with a circuit produced a flow of electricity -- and because electric flows could establish and collapse inductances when they were started and stopped, the equations viewed light as the proportional interaction of collapsing magnetic and electric fields. The light wave was propagated by the force of the two as they alternately collapsed (with the energy producing the establishment, like the energy of the alpha, beta and gamma radiation, or of the neutron, simply assumed.)

By adapting Maxwell's equation to explain light, British science showed itself to be blindly committed to light as a wave, damn the medium!

However, if the radiation from the uranium and the cathode ray tube were considered to be a wave phenomena and not particles, they could not be used to bombard matter. If they had no physical existence, then they could not be used in the service of Newton's Third Law of Motion to cause reactions in matter or react to collisions with matter. Without collisions, no mathematical calculations could be made using Newtons Third Law of Motion, and the nature of matter could not be probed.

Because Empirical Science needed particles to explain reactions, the line was drawn *through* the electron, it being considered both a beta particle and a moving charge, with the alpha particle on one side and the gamma radiation on the other.

By showing that beta rays were deflected by magnetic fields, the British were able to satisfy themselves and the world that the beta rays were different from light, laying to rest the lie that the First World War was fought over the issue. As a result, Newton's Third Law of Motion was established as the sole method of probing the secrets of the atom. The corollary to the gamma radiation, theoretically massless, was an electron with some mass and the less penetrating alpha particle with more mass.

Because the billiard analogy derived from Newton's Third Law of Motion now controlled the way the structure of matter was to be probed, the billiard ball with the most mass, the alpha particle, was selected to do the probing.

If alpha particles were to be the cue balls, then a source of alpha particles had to be obtained.

A RELIABLE SOURCE OF ALPHA PARTICLES

If instead of thinking about a match being immersed in the expanding sphere of emissions from another match, its atoms becoming more active as the field replacement increases, we think about a fixed field in which a range of elements exist, we can see that lining up the elements by weight creates a boundary for an element's existence at the upper level of the field. This boundary, which can be analogized to the point in the field at which the elements that make up the match head start to emit, is dictated by the existing field at the surface of the Earth. Instead of the field, the temperature, being raised until it affects the stability of a particular element, the field limits the elements that can stay together, that can exist in that field. Atoms above a certain weight, with more than a certain number of units in their nuclei, are unstable and simply cannot continue to exist in this existing temperature without breaking down into simpler elements. On a molecular basis the point is the melting point, while with respect to an atom's orbital electrons, the point is the ignition point.

Our concepts concerning the existence of matter were unfortunately shaped by, of all people, Darwin. Early in the nineteenth century, Prout had posited that all atomic numbers and thus elements were multiples of the atomic weight of hydrogen. After Darwin, Lockyer recast Prout's proposition in terms of evolution, with all elements being *descended* from hydrogen. This ridiculous notion has been dressed up in quantum folderol so that today we have the heavier elements being manufactured out of helium located in the centers of the stars, a fantasy given fairy tale life by the skeptical Sagan's oft asserted claim that we are made of star dust.

The measurable reality is that, instead of heavier elements being formed at higher temperatures, elements break down when heat is applied to them, with the highest level of heat needed to break down the lightest element.

That's why engineers use heat to crack petroleum!

The converse, that elements that are too heavy to exist in a certain field can't exist in that field is measurable by the fact that radioactive elements need to be in molecular balance with lighter, more stable elements to survive in the existing field. It is only after they are extracted from the molecular structures that give them stability that they begin to break down, their breakdown measurable by the activity, the emissions they produce, as they seek to become stable in the given field.

Thus, the lighter elements that makes up pitchblende yield uranium, a heavy element at the boundary, with the highest number of units in the nuclei of the atoms that make it up. The uranium has remained stable in the pitchblende in much the same way that zinc remains stable at temperatures over its melting point of 420C when it is in molecular balance with copper to make up brass. The molecular combination allows it to remain stable in a much higher field than it could otherwise exist. When uranium is isolated from pitchblende, exposing it to the existing field, it begins to emit in much the same manner that the match begins to emit, with the difference that the match has to be ignited at a point so far above the existing field of the elements that make it up that it burns furiously to regain balance with the surrounding field (an atomic analogy being to the actual ignition of the units in the nucleus of the atom itself by nuclear fields, the highest possible field, resulting in nuclear explosions.)

A less unstable, and thus more endurable element, radium, obtained from uranium by the Curies, was very effective in producing cancer, so it came into widespread use in treating cancer patients. A by-product of radium treatment was an inert gas, radon, and radon was a particularly good source of alpha particles.

Because radon was easy to work with, was plentiful as a by-product of radiation therapy, did not require a power source and produced an abundance of measurable alpha particles, it became the tool of choice to probe the secrets of the atom.

It was the scattering of alpha particles by gold foil that led Rutherford to deduce the nucleus of the atom.

A QUESTION OF RECOIL

When the nucleus of the atom was deduced, there was no neutron. Alpha particles, helium nuclei with an atomic weight of four, were thought to be made up of four protons, the positive charge Rutherford theorized for the center of the atom to hold the negative electrons in orbit.

Rutherford's experiment with gold foil involved directing a flow of alpha particles at the foil so that they would bounce off at an angle. A glass pane coated with zinc sulfide was placed to catch the bouncing alpha particles. Because the alpha particles were bouncing all over the place, the angle produced by their path was bisected with a lead shield to ensure that they weren't hitting the glass pane directly from the source.

The thought process underlying the experiment is simple. A particle, the alpha particle, is emitted from radon. It travels as a result of its assumed momentum until it hits something. When it hits something, it has an equal and opposite reaction, Newton's Third Law of Motion, and it bounces off of whatever it hits. The reason the alpha particles were scattering was that they were bouncing off of widely separated atoms, revealing the space that separates the atoms and thus the nuclei of the atoms.

Because the alpha particles could be used like projectiles, they could be used to probe the nature of matter that was otherwise too small for measurement. A subatomic picture could be created similar to the way that X-rays produced pictures of bones. However, where X-rays produced a picture by having their straight-line motion blocked from affecting a detector, producing a picture on photographic film by exception, the picture produced by the alpha particles had to be deduced applying Newton's Third Law of Motion to their recoil. If an alpha particle was supposed to be going in one direction and it produced a spark somewhere else, then it had to have hit something and bounced off or at least been deflected.

We can see what happens to billiard balls (or light bouncing off a mirror, for that matter.) We can measure the angles because we know that for every action there has to be and equal and opposite reaction.

We can't see subatomic particles, we can only see sparks, but when the sparks appear out of alignment with their source, we can draw the particle's path from the source and the particle's path to the spark, and locate what it bounced off of by where the two paths intersect.

Using billiards, the analogy is extended to recoil particles. If one particle is emitted from the source and another particle produces a spark, then the second particle is considered to be a recoil particle. When hydrogen nuclei were identified, their recoil was more energetic than alpha particles. Thus, if an alpha particle theoretically hit a hydrogen atom, the alpha particle, being a helium nucleus with four times the mass of the hydrogen nucleus, mathematically produced a much more energetic recoil in the hydrogen nucleus than if the alpha particle hit a helium nucleus of equal mass.

The billiard analogy, however, is deceptive. Subatomic particles clearly can be blocked, leaving an image of the blocked area on photographic plates and they can clearly be scattered, producing a picture of sorts of what is scattering them.

But what is the evidence that they can knock particles about so that the particles being detected are recoil particles?

The proof is in the notion that, because recoil particles have different energies, their momentum is being obtained from the alpha particles.

Applying Newton's Third Law of Motion to the different distances the particles passed through layers of metal foil, and using the different distances as an indication of the particle's energy, the energy was assumed to be transferred to the particle by its collision with the alpha particle. The alpha particle's momentum was transferred to the helium or hydrogen nuclei, with the helium or hydrogen nuclei recoiling as a result of the impact, and the hydrogen nuclei recoiling further because of the mass differential.

But, without a source for the motion of the alpha particles, the application of Newton's Third Law of Motion was actually being used in reverse, to compute the momentum of the alpha particle by assuming its energy was being transferred to the recoil particles whose relative energies were measured by the number of metal foils they could traverse.

This mirror misanalogy is not uncommon when balancing equations is the tool used to define reality. A mirror misanalogy is a reversal of hypothesis. One side of an equation has to contain a mathematical representation of the hypothesis and the other the mathematical computation of the predicted fact. When the hypothesis is found not to predict the outcome, it is altered until it does. Soon the outcome is predicting the hypothesis, the most notorious example being weather models. In mirror misanalogies, the failure of the hypothesis to predict the outcome results in the outcome being used to compute the hypothesis, which then becomes fact. The granddaddy of all mirror misanalogies is Newton's matter/gravity into mass/gravity, where because matter could not be used to predict the inverse square law, the inverse square law is used to compute otherwise unmeasurable mass.

The evidence that the particles were recoiling as a result of being hit by an alpha particle was the fact that different particles traveled different distances through intervening metal foils, an unfounded assumption without knowing the source of the alpha particle's momentum.

By ignoring the source of the alpha particle's motion by computing its force with reference to the reaction it was assumed to produce, an alternate source for the movement of the "recoil" nuclei was also ignored because the same force moving the alpha particle could also be moving the "recoil" nuclei, meaning the "recoil" nuclei would not be recoiling.

THE SOURCE OF THE ALPHA PARTICLE'S MOTION

When an unlit match is immersed in the field of a burning match, it is fairly easy to see the expanding sphere of heat and light. The elements that make up the match head are stable for temperatures at some level above the normal field of the Earth to keep them from igniting when they shouldn't. But eventually the unlit match moves deeply enough into a field to ignite, and when it does, the orbital clouds of the atoms that make up the elements of the match head burst out in the movement of electrons which we see as structured light and feel as heat.

Empirical Science, of course, says that light is made up of insubstantial photons and that because light is sensed with the eye and heat with other senses, the two are separate and distinct, but then Empirical Science bases its knowledge on naming things rather than understanding process.

An electron with the property of motion, the property of moving at the speed of light, solves a lot of problems. One of the reasons that light was considered an imponderable was that its motion couldn't be explained. If motion is assumed for the electron, then different levels of motion, different energy levels don't have to be assumed for alpha, beta and gamma radiation.

Energy becomes the release of the particles that are held together in matter, motion, of course, being the definition of energy.

With motion assumed for electrons, the source of motion for the alpha particles can be found in the production of the beta rays, which are electrons. If electrons are considered to be in motion unless they are restrained, then when elements are composed of boundary atoms that are breaking down in the existing field, the process of breaking down releases electrons in the form of beta radiation. As the breaking down process means that the atoms that make up the element are themselves breaking down, then individual units, or stable combinations of these units are also a by-product of the breakdown, the nuclear disintegration.

These units, formed into what we measure as alpha particles, helium nuclei, do not themselves have motion. The only motion they obtain is by being caught up in the departure of the electrons, the beta radiation, with the electrons imparting part of their motion to the alpha particles.

Thus, the only *momentum* the alpha particles have is the momentum obtained from being caught up in the storm produced by the departing electrons.

If the alpha particles obtain their momentum from the beta radiation, then so would the particles that were considered to be targets of the alpha particles. *Instead of particles striking the detector as a result of the recoil from being hit by an alpha particle, the particles hit the detector as a result of the momentum they obtain from the beta radiation.*

But the recoil nuclei are found to have different levels of energy depending on whether they are helium or hydrogen nuclei. If the lighter hydrogen nuclei move through more layers of foil than the heavier helium nuclei, doesn't that prove that they are in fact recoiling?

The quantity and quality of the scintillations produced in the detector control our picture of what is occurring when various gases are exposed to radiation.

What exactly is a detector detecting and what is actually happening when, as in the case of a particle "absorbed" by the metal foils, a detector ceases to detect what it is designed to detect?

And as an added fillip, what happens when a detector detects something it wasn't designed to detect?

REGISTERING UNSEEN PRESENCES FOR OUR SENSES

When we are dealing with the subatomic world, we can't use our senses to produce a picture of what is going on and we can't create rulers to define reality because our rulers are made up of what we are trying to measure.

We can only produce secondary measuring devices, detectors which, like mediums hovering over crystal balls, register unseen presences for our senses.

Because detectors become our senses, it is of the utmost importance that we understand how detectors work, more important even than knowing what they detect, because if we don't know how a detector works, then we have no way of knowing what it is detecting!

The process of Empirical Science, however, is to make up something that needs detection and then create a detector that will detect it!

The Geiger Counter is a prime example of a detector produced to detect what it was designed to detect, created as it was to produce an electronic depiction of the subjective sparks on panes of zinc sulfide coated glass. It is a sort of reverse cathode ray tube, an anode wire inserted into the center of a cylindrical cathode placed in a tube filled with gas. Detectable events ionize, which is to say, field replace, the atoms of gas, releasing electrons which complete the circuit, allowing a current to reach the counter, in the case of the Geiger Counter, a noisemaker that goes "click." When enough detectable events release enough electrons to complete the circuit between the anode and cathode, a click ensues, and the clicks are used to represent whatever it is that is being detected.

Ignoring the fact that Empirical Science doesn't know what electricity is, what causes it to flow or even why it needs a circuit to flow, any detector it designed to detect by representing clicks couldn't detect anything if the detector was set in a manner that whatever it was detecting always completed the circuit. In such a case, the detector would just be a complete circuit.

Therefore the detector had to be calibrated in such a way that it was an incomplete circuit before it begins to detect. This means that whatever is being detected is just that portion of the population of detectable events that exceeds the population the detector has been calibrated to detect.

The detector can't be used to detect alpha particles until it can be set to separate them out from the background radiation, the beta radiation. If an experiment involves the measurement of alpha particles, and those alpha particles are being accompanied by the electron flows that make up the beta rays, then the detector has to be set to eliminate the background beta rays. When radon emits detectable alpha particles, it is also emitting beta radiation that the detector has been calibrated to ignore. *Once the detection of the beta rays has been eliminated, however, the beta rays are eliminated from awareness.*

In fact, they are still there. They are just undetected!

As in the case of the creation of the nonexistent neutron which masks the nature of energy by assuming momentum, the reality that the detector eliminates might well be more important than the reality that the detector does detect. As empiricists were only interested in experiments requiring a reaction, they focused on alpha particles rather than the beta radiation which produces the very field in which the various atoms, like an unlit match in the flame of a lit match, were being immersed!

The question of whether the detector is detecting recoil particles concerns detectable events that cease to be detectable. If the recoil nuclei are receiving energy from alpha particles rather than being swept along by the beta radiation, then their ability to spark the detector depends solely on the energy they bring to their passage through the metal foils. However, the nuclei may well just be losing their ability to spark the detector.

They may be reaching the detector, they just may no longer be capable of detection!

If this is the case, then their detection would have nothing to do with the energy they received and there would be no need to create an energy differential based on the recoil analogy to explain the fact that different particles travel different distances through the layers of foil.

Whether or not particles are sparking the detector depends on *how* the particles spark the zinc sulfide atoms which in turn requires knowledge about how the atom is constructed, what it is made up of, what the alpha and beta rays consist of and how one striking the other produces light.

Using the analogy of the two matches, we can see that the alpha and beta rays, both of which have an exposed charge, are field replacing the atoms that make up the zinc sulfide as they pass through the zinc sulfide molecules. The atoms that make up the molecules give up their orbiting electrons in sufficient numbers to produce light as a result of field replacement. This explanation is consistent with the size of the scintillations, with the alpha rays producing brighter strikes than the beta rays.

If the particles that pass by the zinc sulfide molecules have to have an exposed charge in order to spark the molecules, then the helium and hydrogen nuclei would cause the molecules to spark only so long as they had exposed charges. They would have exposed charges only so long as they didn't have orbiting clouds of electrons.

When the helium nuclei are separated from the radon nuclei and swept along by the electron flows, they have no orbiting clouds of electrons. However, the nuclei begin to collect orbital clouds as soon as they are separated. If helium nuclei are traveling in a vacuum, there are no atoms available to provide a source for the necessary electrons, the only electrons being those occupied in sweeping the nuclei along. However, when the nuclei begin to pass through a gas or a medium such as air, or even the metal foils, they begin to collect electrons to fill in their orbital clouds of electrons from the orbital clouds of the atoms available in the environment.

Completing their orbital clouds transforms them from helium nuclei, alpha particles without orbital clouds of electrons, into helium atoms, alpha particles with orbital clouds of electrons, *and as such, undetectable.*

They may well hit the detector, they are just incapable of producing a spark.

The beta rays, on the other hand, are themselves flows of electrons and will continue to spark the detector so long as the flows can maintain their integrity.

When the number of foils the hydrogen nucleus moves through is compared to the number of foils the helium nucleus can pass through, the hydrogen nucleus is considered more energetic. However, the hydrogen nucleus has an orbital cloud twice as large as the orbital cloud of the helium nucleus. *It takes a proportionally longer time to complete its orbital cloud than the helium nucleus and thus travels proportionally further through the intervening metal foils before it does so.*

Both nuclei are being swept along by the beta radiation. The helium nucleus just complete its orbital cloud more rapidly than the hydrogen nuclei, and in doing so becomes undetectable sooner.

The made-up recoil, the transfer of momentum created applying Newton's Third Law of Motion to subatomic particles becomes a labored explanation for a simple reality!

One of the most pervasive misperceptions in Modern Science, the belief that a detector can validate a theory by detecting a particle the theory predicts, is evidenced by the cold fusion imbroglio. The process of fusion was conceptualized and then radiation produced by the conceptual process of fusion predicted. Anything predicted will eventually be found, it only takes enough patience, money and the ingenuity to produce the right detector to find it, although, to be truthful, nuclear radiation detectors are a bit too ingenuous, being things of sophistry, their theoretical underpinnings so tenuous as to make them candidates for the Empirical Hall of Fame next to the current concept of the aether as a medium for gravity waves.

Producing a specific detector to detect a specific type of imaginary radiation precludes knowledge of what other things the detector might also detect. If a detector that was accidentally left on detects when there were no experiments around producing the particles which the detector was designed to detect, the detection could easily be dismissed as aberrant. But when the detector is brought into play for experiments where it has no applicability and it still detects, what it detects is mistaken for what it was designed to detect and all hell breaks loose.

The controversy over cold fusion is not a conflict over whether cold fusion is feasible, hell, hot fusion isn't even feasible, its search having been carried on for decades in deference to the age old adage, let milking cows milk. The controversy is over the exposure of the integrity, or rather lack of integrity of the fusion detector itself and thus the quantum theory the detector purports to validate.

If a detector detects something that it wasn't designed to detect, then what it was designed to detect is called into question!

SPLITTING THE ATOM

The experimental apparatus designed to analyze the theoretical recoils involved in alpha particles striking the atoms of various types of gas was an airtight chamber with a radon source placed at one end and a detector, to record scintillations, placed at the other end. Alpha particles from the radon source entered the chamber, supposedly collided with atoms of gas in the chamber, the nuclei of the gas atoms theoretically recoiled as a result of Newton's Third Law of Motion, and the results were recorded as scintillations on the detector.

After producing systemized measurements by bombarding hydrogen and helium with alpha particles emitted by the radon, the view of the hydrogen nucleus as Rutherford's proton began to emerge. Once the hydrogen scintillations were identified, however, they started showing up after the chamber had been evacuated!

At first, it was thought that the hydrogen scintillations were protons being emitted by the radioactive source. Radioactive matter, however, had never been observed to emit protons and it was decided that the hydrogen nuclei had to be coming from somewhere else.

A search for their source began.

A chamber evacuated of air that still has something in it implies that that "something" is air. Air can be dry or humid and is made up of nitrogen, oxygen and carbon dioxide. If it is humid, it also contains atoms of hydrogen and oxygen formed into water molecules. To find out which was responsible for the scintillations, the chamber was filled with each of the suspected gases in turn.

Carbon dioxide, oxygen and humidity didn't materially increase the hydrogen scintillations. Using Newton's Third Law of Motion to explain the result, the experimenters concluded that the alpha particles were being slowed down and thus did not have the energy to *affect* the atoms of gas.

Because some of the atoms of gas, the oxygen and nitrogen, were not hydrogen atoms and were too heavy to theoretically recoil when struck by helium nuclei, what were these experiments really designed to show?

As with most experiments, the desired conclusion was known in advance. Under normal theoretical considerations, the only situation in which the hydrogen scintillations could be expected to increase would be when the air was humid because that was the only situation in which hydrogen atoms would be available. The experiments, however, weren't designed to knock hydrogen atoms out of water molecules.

The thought process underlying the observation that hydrogen scintillations were being produced from an evacuated chamber was that stray atoms of nitrogen or oxygen had stayed behind. The alpha bombardment was thought to be knocking protons, hydrogen nuclei, out of the nitrogen or oxygen nuclei!

When carbon dioxide, oxygen and even humid air was found to reduce the number of scintillations, the experimenters looked at the reduction as the result of the alpha particles losing their energy by their passage through the gas. However, an environment with plentiful oxygen would provide hydrogen nuclei with a source of electrons, which, with their higher charge, they could use to fill in their orbital clouds, making them undetectable when they reached the detector.

When dry air was found to increase the scintillations, the experimenters viewed the energy of the alpha particles as being undiminished when in fact, there were just fewer oxygen atoms around for the hydrogen nuclei to use as a source of electrons to complete their orbital clouds leaving more with detectable charges to reach the detector and produce scintillations.

The hydrogen scintillations that resulted when the chamber was filled with nitrogen gas, however, allowed the experimenters to use the precise numbering system from the periodic table to explain where the hydrogen scintillations were coming from. The helium nucleus with four protons was slamming into the nitrogen atom which, with an atomic weight of fourteen, had fourteen protons. Seventeen of the protons fused (transmuted) into the oxygen isotope O17, leaving one proton, the hydrogen nucleus, to be picked up by the detector.

In reality, the only place that the hydrogen nuclei can come from is the nuclei of atoms, be the atoms the radioactive material or the material placed in the chamber and only then by release. The way the hydrogen nuclei are released is through the field replacement occurring from the field of the environment acting upon the units in the atom. With radioactive elements, the number of units in the nuclei is greater than can stay together in the field in which they exist, the field of temperature at the surface of the Earth. It releases units packaged into helium nuclei. With the atoms in the chamber, *the field is the field produced by the beta emissions from the radioactive material, the very radiation eliminated from the detector's range.* Just as one match will burst into flame when it is moved too close to another burning match, so will the nuclei of the nitrogen atoms begin to lose units packaged into the lighter hydrogen nuclei in the higher field produced by the beta emissions from the radioactive source!

Alpha particles were not knocking a hydrogen nucleus out of a nitrogen nucleus, the beta radiation was replacing the charge holding the units of the nitrogen nuclei together, with some of those units being released as hydrogen nuclei and, swept along by the beta radiation, reaching the detector before they could obtain orbital clouds of electrons.

The extent to which the hydrogen nuclei could be detected depended not on the momentum of the alpha particle but rather on whether they could maintain an exposed charge, which, in turn, depended on the number of available electrons in the environment!

Why doesn't the radioactive element breakdown into hydrogen nuclei?

A QUESTION OF EMISSIONS

Describing alpha, beta and gamma rays as helium nuclei, electrons and electromagnetic radiation in that order is not very informative.

What exactly is being emitted from the radioactive source?

To understand what is going on, we need a description of what is going on, of how these emissions are tied to the nuclei of the radioactive atoms.

Uranium undisturbed in pitchblende deposits is not radioactive although the assumption that it is decaying and producing vast deposits of radon gas beneath the surface of the Earth has sold a lot of radon detectors over the years. Uranium in pitchblende is in molecular balance with the field, the temperature, in which it exists like the elements in alloys are in molecular balance with respect to the alloy's melting point rather than the melting points of the individual elements. The field is the temperature at the surface of the Earth. When the pitchblende is mined, its physical surfaces are exposed to molecular imbalance, exposing isolated uranium atoms which begin to emit, causing the pitchblende itself to break down, further isolating uranium atoms to produce additional radiation.

With motion being a property of the particles that make up the atoms, and with the magnetic assumption of polarity removed from the atom so that the physical law that likes repel can be repealed for the subatomic world, what is being emitted is either electrons or units made up of electrons, with the electrons regaining their motion upon being released, motion being the definition of energy, and the units being swept along to the extent that they cannot be held together in the nuclei of the atoms because the field is too strong for them to remain together.

Why are helium nuclei, rather than hydrogen nuclei, produced as the nuclei of the atoms that make up radioactive matter break down to the extent necessary to reach balance with the existing field?

Because hydrogen atoms cannot exist in the existing field, the temperature at the surface of the Earth!

If hydrogen atoms could exist in the existing field, we wouldn't have water and we wouldn't have the life it produces sitting around asking questions about hydrogen atoms. Hydrogen combines with other elements for much the same reason uranium needs to be in combination with light elements if it is to survive in the existing field. Hydrogen seeks balance with other elements so that the resulting molecules will be in balance with the existing field.

Keeping in mind that we are dealing with a nucleus containing billions of units rather than a single proton, the units that come off of the nuclei of atoms immersed in a field that is stronger than will allow them to exist have to come off in a package containing enough units to maintain its balance with the existing field. Alpha particles in the form of helium nuclei are actually made up of the lowest number of units that can exist in the environment in which the radioactive atoms are breaking down, and they are being separated from the nuclei of the radioactive atoms because the field is too great for the units of the radioactive nuclei to stay together and maintain balance.

The hydrogen nuclei on the other hand, are not separated from a nucleus by the field that exists at the surface of the Earth. *They require the much stronger field produced by the beta radiation in the chamber itself which breaks out the units in smaller packages than can exist in the normal field.*

The beta rays sweep the alpha particles and hydrogen nuclei along after they are separated, the beta rays being the electrons from the clouds of orbiting electrons of the radioactive atoms which are being released by exposure to the field and which are regaining their motion, the release occurring without frequency because combustion is not occurring.

Which leaves the gamma radiation.

The gamma radiation is emitted in a manner that makes it more penetrating than the beta radiation. The distance the two travel through a barrier is what is used to distinguish the two. However, if the relative distances they travel are the result of the same particle being released with different stores of energy, the measurements can be attributed to two different methods of release rather than to two different forms of radiation.

The beta radiation is made up of flows of electrons released from orbital clouds *while the gamma radiation is made up of electrons released from the units themselves, the units that make up the nucleus and which are in turn made up of the same particles, electrons.*

As the field builds up *within* the radioactive nucleus from the emissions, which move in all directions and thus are moving toward the center as well as out, creating a strong nuclear field in the units, some of the units themselves come apart into their constituent electrons.

The orbiting electrons were already expending energy orbiting the nuclei so that when they regain their motion, they regain only a portion of their total energy.

The electrons that make up the gamma radiation, however, were totally immobilized in the units that make up the nucleus, and when they regain their motion, they regain their total energy. Here, with momentum being a product of acceleration, it really is a question of which is more energetic, with the electrons released from stable positions in the units regaining far more momentum than those being released from orbits where they are already in motion.

IMAGINE THE ASTONISHMENT!

Because gamma radiation was so much more energetic than alpha particles, Bothe began experiments using gamma radiation to determine if elements heavier than nitrogen could be split in the same way that the alpha particle was thought to be splitting the nitrogen atom. Mounting polonium, a source of alpha particles comparatively free of gamma radiation, he began to bombard elements in search of one that produced large amounts of gamma radiation. Elements such as boron and aluminum produced gamma radiation, but also disintegrated, producing hydrogen nuclei in the process. These elements couldn't be used because the hydrogen nuclei they produced could not later be separated from the hydrogen nuclei that were supposed to be knocked out by the gamma radiation. When lithium and beryllium were bombarded by the alpha particles, however, they emitted gamma radiation without emitting hydrogen nuclei.

Beryllium, in fact, emitted one heck of a lot of gamma radiation, estimated at ten times the amount of energy received from the alpha bombardment.

Chadwick, who frowned on Bothe's experiments to start with because gamma radiation was not a particle and shouldn't be able to knock anything out of anything, looked askance at this observation because it violated the law dictating that the energy produced could not exceed the energy source.

Bothe's result, however, wouldn't be a surprise to anyone who was looking at the beta radiation field replacing the beryllium nuclei. After all, if we take ten unlit matches and move them into the field of a lit match, the ten will ignite just as easily as the one and produce ten times the energy that one match produces.

The Joliot-Curies, however, oblivious to Chadwick's theoretical disagreement concerning Bothe's results, used the gamma radiation from beryllium as a source of bombardment. When they placed paraffin between the gamma radiation source and an ionization chamber, they found that the bombardment resulted in an increase in the current in the ionization chamber, an observation similar to increased clicks from a Geiger Counter. Because they believed that the detector was recording hydrogen recoils, they could estimate the energy produced by the gamma radiation at three times the energy created by the gamma radiation. They published their results, using the energy of the gamma radiation from the beryllium to account for the detection of the hydrogen nuclei.

The Joliot-Curies' results wouldn't be a surprise to anyone who was looking at the gamma electron radiation bathing the bonds of the hydrocarbon molecules. The orbital clouds holding these atoms together regain their motion, the molecules fall apart, and the individual nuclei are swept along, with the carbon nuclei already in balance and thus undetectable, and the hydrogen nuclei sparking the detector or at least flicking its needle.

Chadwick, however, objected strenuously. He took some measurements of the radiation being produced by the beryllium and found that the radiation was more energetic when it was traveling in the direction of the alpha radiation than when it was traveling in the direction opposite the alpha radiation.

Having concluded that gamma radiation was one of Maxwell's electromagnetic waves, he claimed that, like light, it should propagate evenly in all directions.

He therefore claimed that because the radiation being emitted by the beryllium was not being emitted uniformly in all directions, it was not an electromagnetic wave and therefore could not be gamma radiation.

Anyone who remembered that the polonium source was emitting beta radiation would have realized that gamma radiation emitted in the direction of the beta radiation would lose some of its energy when compared to gamma radiation that was being emitted in the direction of the beta radiation.

Chadwick was therefore arguing with limited awareness!

But that didn't stop him from arguing!

Duplicating the Joliot-Curies' experiments, with the addition of a detector that registered the results by deflecting the needle of an oscilloscope which turned the sparks into a mechanical event that could be calibrated, Chadwick found that the "beryllium" radiation was knocking hydrogen nuclei out of the paraffin.

When he substituted heavier elements for the paraffin, he found that the number of "recoils," the number of hydrogen nuclei registering on the detector, remained the same as the weight of the element increased. The same beryllium radiation knocked out hydrogen atoms in elements with different weights.

This wouldn't be a surprise to anyone who was looking at the beta electron radiation bathing the various heavy atoms.

But if anyone were thinking in terms of Newton's Third Law of Motion, as was Chadwick, that for every action there has to be an equal and opposite reaction, then the same energy knocking hydrogen nuclei out of increasingly heavy atoms would be astonishing.

Why?

It would violate the principle of energy conservation, that energy can neither be created nor destroyed, the momentum of the beryllium radiation being unchangeable.

Gamma radiation did not exist between its source and its effect, it had no mass, or if it did, it had very little of it, the Compton Effect showing that it could merely deflect electrons. Therefore, *saying that gamma radiation could strike a hydrogen nuclei, a proton, and cause it to recoil would be like saying a grain of sand could knock the cue ball off the billiard table*!

THE BIRTH OF THE NONEXISTENT NEUTRON

The Joliot-Curies were about as elegant an authority as could possibly exist in the Empirical Science of the time.

If Chadwick could knock the Joliot-Curies off, their authority would recoil to his benefit!

Chadwick had carefully crafted his experiment to make the Joliot-Curies' claim that the gamma radiation bombardment of the paraffin was knocking the massive hydrogen nuclei into the detectors appear frivolous so that the Joliot-Curies would look silly.

Because very little about Empirical Science has to do with describing what is actually happening in reality, authority is the only basis for making or accepting any statement concerning physical reality. Authority in turn is derived from consensus agreement as to which explanations, which concepts, actually, which beliefs, will be accepted as proven correct.

One of the best ways to undermine authority is ridicule, and in fact, Empirical Science, and to a greater extent, Modern Science, which bases its measurements on accepted beliefs rather than on measurements in physical reality, uses ridicule to dispose of any beliefs that oppose, and thus endanger the accepted beliefs of the consensus.

If something can be dismissed as laughable then it doesn't have to be entertained as a serious challenge to accepted beliefs. This tried and true method is as old as religion, and science is a more rigorous belief system than most religions.

Between Chadwick and the Joliot-Curies, it was an "either/or" situation in spite of the fact that explanations concerning physical reality are never "either/or" situations.

No explanation for any physical effect can ever be more than a mere interpretation of fact!

What made the Chadwick/Joliot-Curie controversy an "either/or" situation, however, was the made-up belief that energy can neither be created nor destroyed. This belief was created to limit reality so that mathematical systems could claim they actually described reality. Applying such a belief to physical reality without an adequate definition of energy, without knowing what it is meant to create or to destroy energy, or even what energy is, is asininity at its most pervasive.

But things like the conservation of energy are beyond textbook learning, they are basic to the mind, like gravity as a property of mass, authority unto themselves, so that to argue against Chadwick's use of Newton's Third Law of Motion, itself basic to the mind, by claiming that gamma radiation could "knock" the protons out of the paraffin would be arguing against the conservation of energy, committing suicide instead of retiring from the field of battle, disgraced, but alive. Graceful disgrace, after all, invites sympathy and Nobels, which rapidly followed.

Chadwick thus had a clear field to come up with his neutron, something that he had long cherished displaying. "It is evident" he said, "that we must either relinquish the application of the conversation of energy and momentum *in these collisions* or adopt another hypothesis about the nature of radiation."

If an attempt had been made to deal with the nature of radiation, not adopt another hypotheses, for there was no hypothesis about radiation to start with, but rather an attempt to come to terms with what energy is, Chadwick might have uncovered the source of the radiation's momentum. If he had abandoned Newton's Third Law of Motion, he might have realized that what happens when a match is immersed in the field of another match is actually the result of subatomic processes.

But Chadwick was intent on creating his neutron and as a result, with the Joliot-Curies vanquished from the playing field, the neutron became a reality!

The neutron had no charge. That was why it could penetrate the same distance that gamma radiation penetrated. It had a mass equal to the proton. That was why it was able to knock the protons about so that they produced hydrogen deflections in the detector. And to put the neutron in the nucleus of the atom, it was only necessary to convert a proton into a neutron because of the similar mass, the proton's mass being one electron shy of a neutron.

Protons could thus match the number of electrons in orbit to determine charge, protons and neutrons could be added together to produce weight, and the world could rest assured in the knowledge that its comprehension of the basic nature of matter and energy was complete.

THE AFTERMATH OF THE MISCONCEPTION OF THE NONEXISTENT NEUTRON: A WEALTH OF MISCONCEPTIONS

Not recognizing the effect of beta or gamma radiation on atoms, but instead applying both polarity and Newton's Third Law of Motion to the analysis of atomic structure has resulted in an atom with an unimaginably large number of subatomic particles, not many of the same particles, as is the case, but many different particles, so many in fact that the number permitted has had to be limited by international agreement. In order to get around this international limitation (which is actually a limitation on the proliferation of Nobels), the number of particles is increased without increasing the number of names by producing particles that mirror each other because of assumed polarity and the number of particle effects is increased without increasing the number of names by creating particles that can be mixed and matched in different combinations so that different mixtures can produce the same particle with different effects.

It's a mess, and an incomprehensible mess at that, but it is a mess derived from the failure to recognize that energy is the release of the particles that make up matter when that matter is immersed in radiation, requiring instead an equal amount of energy to produce the energy we measure. The equal amount of energy is provided by making up particles and then giving those particles momentum which, because it is assumed, assumes away the source of the energy.

Without the release of particles in a field to explain energy, particles like the neutron keep having to be created to explain the energy by assuming they have momentum, and still more particles have to be created to explain the energy the new particles produce when their energy doesn't balance the energy the new particles produced to account for the unbalanced energy.

It is an endless balancing act, Charlie Chaplin reducing a table to rubble by mis-sawing one leg at a time, only it is our understanding that is being reduced to rubble.

Because matter becomes more stable toward the center of the periodic table, gamma radiation cannot at its normal energy level split atoms with higher atomic weights. Thus, cyclotrons, linear accelerators and superconducting supercolliders are crafted that accelerate charged particles so that they smack against targets at tremendous velocities, literally blasting the atoms of the target into smithereens. Repeating patterns are gleefully picked out from the billions of resulting swirls and tracings created by allowing the trajectories of the various smithereens to expose photographic plates. The patterns are then ingeniously crafted into particles to explain motion and energy. The goal of this puerile process is to find the ultimate particle that explains everything and has resulted in quantum mechanics, an incomprehensible body of pseudo-knowledge created by supposedly sane theorists that claims, among a multitude of other litanous idiocies that the same particle can be in two different places at the same time!

And all that is really happening is that bits and pieces of the units that make up the nuclei of all atoms are flying off depending on the angle and force of impact, and the strength of the background radiation and surviving depending on the size of the piece and the availability of electrons in the environment, until they in turn disintegrate into gamma radiation or recombine into or with atoms.

Ignoring the effect of radiation in the release of energy, and instead creating the neutron out of whole cloth didn't solve the energy equivalency problem because the alpha particles were still producing ten times the amount of energy the alpha particles were bringing in under Newtonian third law computations and the beryllium radiation was still disintegrating atoms. To magnify the matter, Fermi found that beryllium radiation could activate light elements for short periods of time, and when the light elements were placed near a source of hydrogen, the radioactivity was enhanced immeasurably. The unimaginable amount of energy that became theoretically possible as a result of such ruminations was more than conversions from horsepower to steam, or steam to electricity could contemplate.

Looking around for some sort of analogy to correlate the production of nuclear energy, the actual ignition of the units in the nucleus of the atom by a very strong field, with the apparent absence of a source, Meitner's gaze fell on Einstein's great conceptual equation, e=mc2, its use in explaining nuclear energy in turn accepted as validation of Einstein's attempt to describe a universe in which space and time themselves were quantifiable.

$E=mc^2$ is a conceptual equation because each of its terms represents a concept. The "c^2" requires that the speed of light be defined in short units so the number produced becomes conceptually unimaginable, conveying a sense of the massive amounts of energy bound up in matter. If the distance light travels in a second were classified as "1" as most standard units are, then its square would be "1" and the resulting energy minuscule. The "m" mass is an unknown quantity. It is the imaginary matter produced by applying an inverse square measurement and the mass/gravity concept to the movement of matter in space. It includes black holes and dark matter, things that, besides being nonexistent, are unmeasurable. The "e" is, well, energy is a concept all its own, without any comprehensible definition other than motion.

The conceptual meaning of the equation, that energy is made up of the matter that produces it, even opposes the current empirical understanding of matter and energy. The equation says that matter is made up of the energy it can produce, that the units in the nuclei of atoms are made up of electrons, as are the orbiting clouds surrounding those nuclei, and that if all of these electrons were released at once, there would be one hell of a lot of heat and light, and that conversely, matter is the structure produced when the motion of the particles that make it up is immobilized.

However, Modern Science considers heat and light as imponderables, phenomena that do not exist between their source and effect, and as a result, not made up of the matter that gives them off. Modern Science, therefore, while using muddled thinking to justify the production of large amounts of energy in the compression of small amounts of unstable matter, actually does not believe that matter is made up of the energy it gives off when it comes apart, and therefore cannot believe in the equation $e=mc^2$.

The concept of equivalency of matter and energy, however, was sufficient to direct trial and error experimentation to the practical conclusion that compressing unstable atoms with an implosion would result in the sudden release of energy. Thus, even though it wasn't understood that the release was actually the release of the particles that make up the units in the nuclei of the unstable atoms in the form of gamma radiation as a result of the nuclear field replacement produced by the implosion, technology was able to purify sufficient amounts of unstable elements and engineer a method to control and direct the implosion so that the resulting field would be high enough to replace the field holding the units in the nuclei of the individual atoms together and produce a nuclear explosion.

Science is the production of technology from the trial and error manipulation of physical reality.

It is not theory!

Theory is the creation of pictures of how physical reality might work.

Most science, most technology, occurs in spite of modern scientific theory which, being mathematically based, limits physical reality to the allowable terms in equations.

Today, when theory has nothing to do with physical reality, when Empiricists dance around the ceremonial fire blowing smoke and waving mirrors, crying out black holes and dark matter, warped space and compressed time to draw attention away from the simple fact that they can't explain why a ball drops or the sun comes up, theory is quick to claim credit for the advancements of science, denigrating the producing technology in the process, casting Edison's light bulb into the darkness of quantum mechanics which claims exquisite precision for its predictions.

This exquisite precision comes from quantum mechanic's basic dictate that nothing can be known until it is measured, removing the difficulty involved in building specific detectors to verify nonexistent facts and allowing quantum mechanics to claim responsibility for virtually everything, a claim, that at last estimate, includes one third of the momentary value of the world's economy, most of which can be laid at Edison's electrification model.

The real problem is not theorists. There is not enough of them.

The real problem is pseudo-theorists who ossify theory into canon so that the scientist, the engineer, is limited in the exploration of the possible.

Printed in the United States
27876LVS00007B/19

9 780962 776977